Geometric Algebra: An Algebraic System
for Computer Games and Animation

John Vince

Geometric Algebra:
An Algebraic System
for Computer Games
and Animation

 Springer

Prof. John Vince, MTech, PhD, DSc, CEng, FBCS
www.johnvince.co.uk

ISBN 978-1-84882-378-5   e-ISBN 978-1-84882-379-2
DOI 10.1007/978-1-84882-379-2
Springer Dordrecht Heidelberg London New York

British Library Cataloguing in Publication Data
A catalogue record for this book is available from the British Library

Library of Congress Control Number: 2009926270

Cover design: Boekhorst Design BV

Printed on acid-free paper

Springer is part of Springer Science+Business Media (www.springer.com)

This book is affectionately dedicated to my family: Annie, Samantha, Anthony, Genny, Peter, Megan, Mia, Lucie and my dog Monty.

Preface

In my first book on geometric algebra in 2007 the preface described how I had been completely surprised by the existence of geometric algebra, especially after having recently completed a book on vector analysis where it was not even mentioned! So why am I writing a second book on the same subject? Well it's not because I have nothing better to do with my time. There are many more books I have to write before going to the great library in the sky!

When I started writing *Geometric Algebra for Computer Graphics* I knew very little about the subject and had to understand the concepts as I went along—which was extremely difficult. In retrospect, the year spent writing that book was like climbing a mountain, and after completing the chapter on conformal geometry I had effectively reached the summit and, in terms of my understanding, the view was compelling. But having reached the summit I then had to race down with my manuscript and send it to Springer.

In the following weeks it was difficult to forget the previous year's journey. Had I really understood geometric algebra? Had I really laid out the subject in a way that anyone could understand? Such questions bothered me on a daily basis, especially when walking my dog Monty. Such moments gave me the time to reflect upon what was really behind the algebra and what had gone on between Hamilton, Grassmann and Gibbs when the foundations of vector analysis were being established a hundred and fifty years ago.

Back in my office I started to explore vector products from a symbolic standpoint and realized that if two vectors are expanded algebraically, four terms result from two 2D vectors and nine terms from two 3D vectors. Nothing new, or earth shattering. However, if these terms are divided into two sets, they give rise to the inner and outer products:

$$ab = a \cdot b + a \wedge b$$

which is Clifford's original geometric product.

I also found that when such products are expanded in tabular form, and colour is used to highlight the inner and outer product terms, the difference between the two sets became strikingly obvious. I immediately asked Springer for permission to use colour throughout a new book on geometric algebra, which would hopefully would open up the subject to a wider audience.

I continued to apply the same algebraic treatment to vectors, bivectors and trivectors and then discovered that I had been using something called *dyads*, which had been employed by Gibbs in his work on vectors. Far from being disappointed, I continued in the knowledge that I was probably on the right track.

The book's structure emerged without too much effort: The first chapter, which was the last to be written, briefly explores the important role that axioms play in mathematics and how we have struggled during previous centuries to accept non-sensical ideas such as intersecting parallel lines, infinite sets and imaginary numbers. This is to prepare the reader for ideas such as multiplying a line by an area, squaring an area, or adding scalars, lines, areas and volumes to create a multivector. It reminds me of the time I wrote some code to add the shapes of an elephant and seahorse together, or divide a circle by a triangle. Totally non-sensical, but very useful!

The second chapter reviews the products of real algebra, complex numbers and quaternions using the same tables employed later for geometric algebra.

The third chapter is on vector products and reviews the traditional scalar and vector products in tabular form. Dyadics are then introduced and lead onto a description of the outer product in 2D and 3D.

Chapter four introduces the geometric product as the sum of the inner and outer products. Blades are defined and the chapter concludes by exploring the geometric product of various types of vectors.

Having laid the foundations for geometric algebra in the first four chapters, chapter five describes features such as grades, pseudoscalars, multivectors, reversion, inversion, duality and the imaginary and rotational properties of bivectors.

Next, chapter six covers all the possible products between vectors and bivectors in 2D. Similarly, chapter seven covers all the possible products between vectors, bivectors and trivectors in 3D. Tables and colour play an important role in revealing the natural patterns that result from these products.

Chapter eight shows how powerful geometric algebra is when handling reflections and rotations, and at this point we discover that quaternions are a natural feature of geometric algebra.

Chapter nine explores a wide range geometric problems encountered in computer games and computer animation problems. It is far from exhaustive, but provides strategies that can be employed in all sorts of similar problems.

Finally, chapter ten draws the book to a conclusion.

Having written these ten chapters I hope that I have finally found a straightforward way of describing geometric algebra that will enable it to be used by anyone working in computer graphics.

I should say something about the notation employed in the book. Vectors are normally shown in a bold typeface, to distinguish them from scalar quantities. But as virtually every equation references vectors, I have followed Chris Doran and Anthony Lasenby's lead and left them untouched. There is no confusion between vectors and scalars, as you will discover.

I would like to acknowledge that I could not have written this book without the existence of *Geometric Algebra for Physicists* written by Chris Doran and Anthony Lasenby. It provides the most lucid introduction to geometric algebra. Similarly, Michael Crowe's *A History of Vector Analysis* is the best book on the subject.

Once again, I am indebted to Beverley Ford, General Manager, Springer UK, and Helen Desmond, Assistant Editor for Computer Science, for their continual support throughout the development of this manuscript.

I do hope you enjoy reading and discovering something new from this book.

Ringwood   John Vince

Contents

Symbols and notation

Greek symbols

α	alpha
β	beta
χ	chi
δ	delta
ε	epsilon
φ	phi
λ	lambda
θ	theta
Σ	sigma
ψ	psi
ζ	zeta

Mathematical notation

i	$= \sqrt{-1}$
$\in$	member of
$\mathbb{R}$	set of real numbers
$n!$	$= n(n-1)(n-2)\ldots 1$
$\cdot$	inner product (alias scalar product)
$\wedge$	outer product (alias wedge product)
$\times$	vector product (alias cross product)
$\vee$	meet operator
$\parallel$	parallel to
$\perp$	perpendicular to
$\dagger$	dagger (reverse in geometric algebra)
$\sim$	tilde (reverse in geometric algebra)

$*$	asterisk (dual in geometric algebra)
$\hat{a}$	unit vector
$|a|$	magnitude of vector a
$\overrightarrow{AB}$	vector between the points A and B
$\cup$	intersection
exp	exponential
$\triangle ABC$	triangle with vertices ABC

1 Introduction

1.1 Sense and nonsense

Axioms are the building blocks of mathematical theorems and define the rules controlling every potential configuration of numbers, operators and other valid constructs. A system of axioms must be well defined, consistent and should be devoid of any redundancy, but they do *not* have to make sense. To clarify this last remark, I mean that what we understand as being 'sensible' has nothing to do with an axiomatic system. For example, if the square of a number results in a negative number, then so be it. Similarly, why shouldn't parallel lines intersect, and why shouldn't one infinite set be larger than another? The history of mathematics records our struggle to comprehend 'non-sensical' ideas such as intersecting parallel lines, infinite sets and imaginary numbers, and today they are taught as if they had always been part of mathematics.

In the 19th century a number of mathematicians were looking for a mathematical system to describe quantities such as velocity, acceleration and force. Complex numbers were thought to be the key, and Sir William Rowan Hamilton was one such mathematician looking for their 3D equivalent. He thought that as a complex number could be treated as an ordered pair of numbers, i.e. a couple, he reasoned that their 3D equivalent would be in the form of a triplet. However, this was not to be so, and after many years of toil, Hamilton discovered quaternions on October 16, 1843.

A complex number is defined as

$$a + ib \tag{1.1}$$

where

$$i^2 = -1 \tag{1.2}$$

whereas a quaternion is defined as

$$a + ib + jc + kd \tag{1.3}$$

where

$$i^2 = j^2 = k^2 = ijk = -1 \tag{1.4}$$

J. Vince, *Geometric Algebra: An Algebraic System for Computer Games and Animation*,
© Springer-Verlag London Limited 2009

with the rules connecting the three imaginaries

$$ij = k \quad jk = i \quad ki = j \tag{1.5}$$

and their non-commuting reversions

$$ji = -k \quad kj = -i \quad ik = -j. \tag{1.6}$$

Unfortunately, the imaginary nature of quaternions were to be their downfall and, after decades of arguing and politics, i, j and k were replaced by $\mathbf{i}, \mathbf{j}$ and $\mathbf{k}$—today's familiar unit basis vectors.

At the time, mathematicians of the day had not seen the wood for the trees, for when Hamilton had initially expanded the algebraic product of two quaternions, he had not known how to resolve the terms ij, jk, ki, ji, kj and ik. It was only on that fateful day, October 16, that the penny dropped and the above rules flashed into his brain. But we now know that if we accept that ij, jk and ki describe a new mathematical object, such rules are unnecessary. In fact, strictly speaking, it is not the imaginary products ij, jk and ki that describe a new mathematical object, it is their vector substitutes: $\mathbf{ij}, \mathbf{jk}$ and $\mathbf{ki}$.

Before Hamilton discovered quaternions Hermann Grassmann had developed his own theory of vectors but was unable to influence the tide of opinion, even though the first application of his notation was employed in a 200-page essay on the theory of tides *Theorie der Ebbe und Flut*!

By the early 20th century, vector analysis had been determined by Josiah Willard Gibbs, who was not an admirer of Hamilton's quaternions. Gibbs recognised that a pure quaternion could be interpreted as a vector $a\mathbf{i} + b\mathbf{j} + c\mathbf{k}$ without any imaginary connotation, and could form the basis of a vectorial system. Two products for vectors emerged: the dot (scalar) product and the cross (vector) product, which could be combined to form the scalar triple product $a \cdot (b \times c)$ and the vector triple product $a \times (b \times c)$. At last, vector analysis had been defined and was understood. But mathematicians were unaware that they had walked up a mathematical *cul de sac*!

Every student of mathematics knows that the cross product has no meaning in 2D, behaves immaculately in 3D, but is ambiguous in higher dimensional spaces. So because of its inherent fussiness, it is not an important mathematical product after all, in spite of its usefulness in resolving 3D geometric problems.

1.2 Geometric algebra

Geometric algebra proposes an alternative vectorial framework where lines, areas, volumes and hyper-volumes are recognised as structures with magnitude and orientation. Oriented lines are represented by vectors, oriented areas by bivectors and oriented volumes by trivectors. Higher dimensional objects are also permitted. At the heart of geometric algebra is the geometric product, which is defined as the sum of the inner and outer products. The inner product is related to the scalar product, and the outer product is related to the cross product. What is so flexible about this approach is that all sorts of products are permitted such as (line × line), (line × area), (area × area), (line × volume), (area × volume), (volume × volume), etc. Furthermore, the cross product has its alias within the algebra as do quaternions; and on top of these powerful features one can add, subtract and even divide such oriented objects.

You are probably wondering how it is possible that such a useful algebra has lain dormant for so many years? Well, through the endeavor of mathematicians such as William Kingdon Clifford and David Hestenes we now have a geometric calculus that is being embraced by the physics community through the work of Anthony Lasenby, Joan Lasenby and Chris Doran. In spite of geometric algebra's struggle to surface, today it does exist and is relatively easy to understand, and I will reveal its axioms and structures in the following chapters.

If at any stage some of the ideas don't make sense, remember the opening remarks at the start of this chapter about common sense. For example, you will discover that oriented areas and volumes are complex quantities, and why shouldn't they be? If it gives rise to a consistent and useful algebra then let it be. You will also discover that it is possible to combine scalars, lines, areas and volumes using multivectors. Such a combination of objects seems rather weird, but if it leads to something useful, then so be it.

Let us now begin to unravel the ideas behind geometric algebra.

2 Products

2.1 Introduction

This is a short chapter and one where we return to some familiar product expansions to identify any useful patterns. Fortunately they do exist and also surface in geometric algebra. So this chapter should be regarded as laying the foundations for geometric algebra.

We begin by investigating real, complex and quaternion products, where each one reveals a consistent pattern that continues throughout the vector products covered in chapter 3.

2.2 Real products

The product of two real expressions such as

$$(\alpha_1 + \alpha_2)(\beta_1 + \beta_2) \qquad [\alpha_i, \beta_i \in \mathbb{R}] \tag{2.1}$$

is effected by multiplying each term in the second expression by every term in the first expression as follows:

$$
\begin{aligned}
(\alpha_1 + \alpha_2)(\beta_1 + \beta_2) &= \alpha_1(\beta_1 + \beta_2) + \alpha_2(\beta_1 + \beta_2) \\
&= \alpha_1\beta_1 + \alpha_1\beta_2 + \alpha_2\beta_1 + \alpha_2\beta_2 \\
&= (\alpha_1\beta_1 + \alpha_2\beta_2) + (\alpha_1\beta_2 + \alpha_2\beta_1).
\end{aligned}
\tag{2.2}
$$

Similarly, products with three terms expand to:

$$
\begin{aligned}
(\alpha_1 + \alpha_2 + \alpha_3)(\beta_1 + \beta_2 + \beta_3) = {}& (\alpha_1\beta_1 + \alpha_2\beta_2 + \alpha_3\beta_3) \\
& + (\alpha_1\beta_2 + \alpha_1\beta_3 + \alpha_2\beta_1 + \alpha_2\beta_3 + \alpha_3\beta_1 + \alpha_3\beta_2).
\end{aligned}
\tag{2.3}
$$

These results reveal two distinct sets of terms that arise whenever such products are expanded. The first set from (2.3):

$$\alpha_1\beta_1 + \alpha_2\beta_2 + \alpha_3\beta_3 \tag{2.4}$$

J. Vince, *Geometric Algebra: An Algebraic System for Computer Games and Animation*,
© Springer-Verlag London Limited 2009

is represented by

$$\sum_{i=1}^{n} \alpha_i \beta_i \tag{2.5}$$

whilst the second set from (2.3):

$$\alpha_1\beta_2 + \alpha_1\beta_3 + \alpha_2\beta_1 + \alpha_2\beta_3 + \alpha_3\beta_1 + \alpha_3\beta_2 \tag{2.6}$$

is represented by

$$\sum_{i=1}^{n}\sum_{j=1}^{n}(1 - \delta_{ij})(\alpha_i\beta_j) \tag{2.7}$$

where δ_{ij} is the Kronecker delta function, defined by

$$\delta_{ij} = \begin{cases} 1 & \text{if } i = j, \\ 0 & \text{if } i \neq j. \end{cases} \tag{2.8}$$

An alternate way to visualize these algebraic expansions (2.2) and (2.3) is shown in Tables 2.1 and 2.2 where we see the inner diagonal forms the first set and the outer terms form the second set. In fact, the terms "inner" and "outer" are quite convenient and will help us later on. For the moment we will refer to these two sets as the inner and outer parts of a product.

As we are dealing with real quantities, both sets of expanded terms are also real. However, this does not hold for imaginary nor vector expressions, which we explore next.

TABLE 2.1   The product $(\alpha_1 + \alpha_2)(\beta_1 + \beta_2)$

$(\alpha_1 + \alpha_2)(\beta_1 + \beta_2)$		
$\times$	β_1	β_2
α_1	$\alpha_1\beta_1$	$\alpha_1\beta_2$
α_2	$\alpha_2\beta_1$	$\alpha_2\beta_2$

TABLE 2.2   The product $(\alpha_1 + \alpha_2 + \alpha_3)(\beta_1 + \beta_2 + \beta_3)$

$(\alpha_1 + \alpha_2 + \alpha_3)(\beta_1 + \beta_2 + \beta_3)$			
$\times$	β_1	β_2	β_3
α_1	$\alpha_1\beta_1$	$\alpha_1\beta_2$	$\alpha_1\beta_3$
α_2	$\alpha_2\beta_1$	$\alpha_2\beta_2$	$\alpha_2\beta_3$
α_3	$\alpha_3\beta_1$	$\alpha_3\beta_2$	$\alpha_3\beta_3$

2.3 Complex products

Given two complex numbers:

$$z_1 = \alpha_1 + i\alpha_2$$

$$z_2 = \beta_1 + i\beta_2 \qquad [\alpha_i, \beta_i \in \mathbb{R}] \qquad (2.9)$$

where

$$i^2 = -1 \qquad (2.10)$$

their product is

$$z_1 z_2 = \alpha_1(\beta_1 + i\beta_2) + i\alpha_2(\beta_1 + i\beta_2)$$

$$= \alpha_1\beta_1 + i\alpha_1\beta_2 + i\alpha_2\beta_1 + i^2\alpha_2\beta_2$$

$$= (\alpha_1\beta_1 - \alpha_2\beta_2) + i(\alpha_1\beta_2 + \alpha_2\beta_1). \qquad (2.11)$$

Once again, we discover two distinct sets: Our so-called *inner* set $\alpha_1\beta_1 - \alpha_2\beta_2$ is a real quantity, whilst the *outer* set $i(\alpha_1\beta_2 + \alpha_2\beta_1)$ is a pure imaginary term. The terms (2.11) are shown in tabular form in Table 2.3.

TABLE 2.3   The product $z_1 z_2$

$z_1 = \alpha_1 + i\alpha_2$ $z_2 = \beta_1 + i\beta_2$		
$z_1 z_2$	β_1	$i\beta_2$
α_1	$\alpha_1\beta_1$	$i\alpha_1\beta_2$
$i\alpha_2$	$i\alpha_2\beta_1$	$-\alpha_2\beta_2$

The conjugate form is often used to convert a complex number into a real quantity, which involves reversing the sign of the imaginary part such that when we form the product of a complex number and its conjugate we have:

$$z_1 = \alpha_1 + i\alpha_2$$

$$z_1^* = \alpha_1 - i\alpha_2$$

$$z_1 z_1^* = (\alpha_1 + i\alpha_2)(\alpha_1 - i\alpha_2)$$

$$= \alpha_1(\alpha_1 - i\alpha_2) + i\alpha_2(\alpha_1 - i\alpha_2)$$

$$= \alpha_1^2 - i\alpha_1\alpha_2 + i\alpha_1\alpha_2 - i^2\alpha_2^2$$

$$= \alpha_1^2 + \alpha_2^2. \qquad (2.12)$$

This time the product contains only the inner set, which is a scalar. Table 2.4 shows (2.12) in tabular form and highlights in red the self-cancelling outer terms.

TABLE 2.4   The product $z_1 z_1^*$

$z_1 = \alpha_1 + i\alpha_2 \quad z_1^* = \alpha_1 - i\alpha_2$		
$z_1 z_1^*$	α_1	$-i\alpha_2$
α_1	α_1^2	$-i\alpha_1\alpha_2$
$i\alpha_2$	$i\alpha_1\alpha_2$	α_2^2

2.4 Quaternion products

A quaternion is the 3D equivalent of a complex number and is represented as follows:

$$q = \alpha_0 + i\alpha_1 + j\alpha_2 + k\alpha_3 \qquad [\alpha_i \in \mathbb{R}] \tag{2.13}$$

with Hamilton's rules:

$$i^2 = j^2 = k^2 = ijk = -1$$

$$ij = k \quad jk = i \quad ki = j \quad ji = -k \quad kj = -i \quad ik = -j. \tag{2.14}$$

The product of two quaternions is expanded as follows:

$$q_1 = \alpha_0 + i\alpha_1 + j\alpha_2 + k\alpha_3$$
$$q_2 = \beta_0 + i\beta_1 + j\beta_2 + k\beta_3$$
$$q_1 q_2 = \alpha_0 \left(\beta_0 + i\beta_1 + j\beta_2 + k\beta_3\right)$$
$$+ i\alpha_1 \left(\beta_0 + i\beta_1 + j\beta_2 + k\beta_3\right)$$
$$+ j\alpha_2 \left(\beta_0 + i\beta_1 + j\beta_2 + k\beta_3\right)$$
$$+ k\alpha_3 \left(\beta_0 + i\beta_1 + j\beta_2 + k\beta_3\right)$$

$$= \alpha_0\beta_0 + i\alpha_0\beta_1 + j\alpha_0\beta_2 + k\alpha_0\beta_3$$
$$+ i\alpha_1\beta_0 + i^2\alpha_1\beta_1 + ij\alpha_1\beta_2 + ik\alpha_1\beta_3$$
$$+ j\alpha_2\beta_0 + ji\alpha_2\beta_1 + j^2\alpha_2\beta_2 + jk\alpha_2\beta_3$$
$$+ k\alpha_3\beta_0 + ki\alpha_3\beta_1 + kj\alpha_3\beta_2 + k^2\alpha_3\beta_3$$

$$= \alpha_0\beta_0 - (\alpha_1\beta_1 + \alpha_2\beta_2 + \alpha_3\beta_3)$$
$$+ i\left(\alpha_0\beta_1 + \alpha_1\beta_0 + \alpha_2\beta_3 - \alpha_3\beta_2\right)$$
$$+ j\left(\alpha_0\beta_2 - \alpha_1\beta_3 + \alpha_2\beta_0 + \alpha_3\beta_1\right)$$
$$+ k\left(\alpha_0\beta_3 + \alpha_1\beta_2 - \alpha_2\beta_1 + \alpha_3\beta_0\right). \tag{2.15}$$

The result still contains two sets in the form of a real inner set:

$$\alpha_0\beta_0 - (\alpha_1\beta_1 + \alpha_2\beta_2 + \alpha_3\beta_3) \tag{2.16}$$

and an imaginary outer set:

$$i \left(\alpha_0 \beta_1 + \alpha_1 \beta_0 + \alpha_2 \beta_3 - \alpha_3 \beta_2 \right)$$
$$+ j \left(\alpha_0 \beta_2 - \alpha_1 \beta_3 + \alpha_2 \beta_0 + \alpha_3 \beta_1 \right)$$
$$+ k \left(\alpha_0 \beta_3 + \alpha_1 \beta_2 - \alpha_2 \beta_1 + \alpha_3 \beta_0 \right). \tag{2.17}$$

Table 2.5 shows (2.17) in tabular form.

TABLE 2.5   The product $q_1 q_2$

$q_1 = \alpha_0 + i\alpha_1 + j\alpha_2 + k\alpha_3$ $q_2 = \beta_0 + i\beta_1 + j\beta_2 + k\beta_3$			
$q_1 q_2$	β_0	$i\beta_1$	$j\beta_2$	$k\beta_3$
α_0	$\alpha_0\beta_0$	$i\alpha_0\beta_1$	$j\alpha_0\beta_2$	$k\alpha_0\beta_3$
$i\alpha_1$	$i\alpha_1\beta_0$	$-\alpha_1\beta_1$	$k\alpha_1\beta_2$	$-j\alpha_1\beta_3$
$j\alpha_2$	$j\alpha_2\beta_0$	$-k\alpha_2\beta_1$	$-\alpha_2\beta_2$	$i\alpha_2\beta_3$
$k\alpha_3$	$k\alpha_3\beta_0$	$j\alpha_3\beta_1$	$-i\alpha_3\beta_2$	$-\alpha_3\beta_3$

The conjugate quaternion reverses the sign of the imaginary part such that when we form the product of a quaternion with its conjugate we have:

$$\left(\alpha_0 + i\alpha_1 + j\alpha_2 + k\alpha_3 \right) \left(\alpha_0 - i\alpha_1 - j\alpha_2 - k\alpha_3 \right)$$
$$= \alpha_0 \left(\alpha_0 - i\alpha_1 - j\alpha_2 - k\alpha_3 \right)$$
$$+ i\alpha_1 \left(\alpha_0 - i\alpha_1 - j\alpha_2 - k\alpha_3 \right)$$
$$+ j\alpha_2 \left(\alpha_0 - i\alpha_1 - j\alpha_2 - k\alpha_3 \right)$$
$$+ k\alpha_3 \left(\alpha_0 - i\alpha_1 - j\alpha_2 - k\alpha_3 \right)$$

$$= \alpha_0^2 - i\alpha_0\alpha_1 - j\alpha_0\alpha_2 - k\alpha_0\alpha_3$$
$$+ i\alpha_0\alpha_1 - i^2\alpha_1^2 - ij\alpha_1\alpha_2 - ik\alpha_1\alpha_3$$
$$+ j\alpha_0\alpha_2 - ji\alpha_1\alpha_2 - j^2\alpha_2^2 - jk\alpha_2\alpha_3$$
$$+ k\alpha_0\alpha_3 - ki\alpha_1\alpha_3 - kj\alpha_2\alpha_3 - k^2\alpha_3^2$$
$$= \alpha_0^2 + \alpha_1^2 + \alpha_2^2 + \alpha_3^2. \tag{2.18}$$

Like the complex conjugate, the product contains only a real inner set. Table 2.6 shows (2.18) in tabular form and highlights in red the self-cancelling outer terms.

TABLE 2.6   The product $q_1 q_1^*$

$q_1 = \alpha_0 + i\alpha_1 + j\alpha_2 + k\alpha_3$ $\quad$ $q_1^* = \alpha_0 - i\alpha_1 - j\alpha_2 - k\alpha_3$			
$q_1 q_1^*$	α_0	$-i\alpha_1$	$-j\alpha_2$	$-k\alpha_3$
α_0	α_0^2	$-i\alpha_0\alpha_1$	$-j\alpha_0\alpha_2$	$-k\alpha_0\alpha_3$
$i\alpha_1$	$i\alpha_0\alpha_1$	α_1^2	$-k\alpha_1\alpha_2$	$j\alpha_1\alpha_3$
$j\alpha_2$	$j\alpha_0\alpha_2$	$k\alpha_1\alpha_2$	α_2^2	$-i\alpha_2\alpha_3$
$k\alpha_3$	$k\alpha_0\alpha_3$	$-j\alpha_1\alpha_3$	$i\alpha_2\alpha_3$	α_3^2

To complete this review of quaternion products, let's consider the product of what Hamilton described as "pure quaternions", i.e. a quaternion with a zero scalar part. Given two pure quaternions:

$$q_1 = i\alpha_1 + j\alpha_2 + k\alpha_3$$

$$q_2 = i\beta_1 + j\beta_2 + k\beta_3$$

$$q_1 q_2 = i\alpha_1 \left(i\beta_1 + j\beta_2 + k\beta_3 \right)$$
$$+ j\alpha_2 \left(i\beta_1 + j\beta_2 + k\beta_3 \right)$$
$$+ k\alpha_3 \left(i\beta_1 + j\beta_2 + k\beta_3 \right)$$

$$= i^2\alpha_1\beta_1 + ij\alpha_1\beta_2 + ik\alpha_1\beta_3$$
$$+ ji\alpha_2\beta_1 + j^2\alpha_2\beta_2 + jk\alpha_2\beta_3$$
$$+ ki\alpha_3\beta_1 + kj\alpha_3\beta_2 + k^2\alpha_3\beta_3$$

$$= -\alpha_1\beta_1 + k\alpha_1\beta_2 - j\alpha_1\beta_3$$
$$- k\alpha_2\beta_1 - \alpha_2\beta_2 + i\alpha_2\beta_3$$
$$+ j\alpha_3\beta_1 - i\alpha_3\beta_2 - \alpha_3\beta_3$$

$$= - \left(\alpha_1\beta_1 + \alpha_2\beta_2 + \alpha_3\beta_3 \right)$$
$$+ i \left(\alpha_2\beta_3 - \alpha_3\beta_2 \right) + j \left(\alpha_3\beta_1 - \alpha_1\beta_3 \right) + k \left(\alpha_1\beta_2 - \alpha_2\beta_1 \right) \tag{2.19}$$

Table 2.7 shows (2.19) in tabular form.

If we substitute the unit basis vectors $\mathbf{i}, \mathbf{j}, \mathbf{k}$ in (2.19) for the imaginaries i, j, k, it is clear that the inner set is the negative of the scalar product of two vectors, whilst the outer set is identical to the

Table 2.7   The product q_1q_2

$q_1 = i\alpha_1 + j\alpha_2 + k\alpha_3$ $q_2 = i\beta_1 + j\beta_2 + k\beta_3$			
q_1q_2	$i\beta_1$	$j\beta_2$	$k\beta_3$
$i\alpha_1$	$-\alpha_1\beta_1$	$k\alpha_1\beta_2$	$-j\alpha_1\beta_3$
$j\alpha_2$	$-k\alpha_2\beta_1$	$-\alpha_2\beta_2$	$i\alpha_2\beta_3$
$k\alpha_3$	$j\alpha_3\beta_1$	$-i\alpha_3\beta_2$	$-\alpha_3\beta_3$

vector (cross) product. In fact, in 1881 Gibbs developed what became modern 3D vector analysis by adopting Hamilton's pure quaternion but without the imaginary connotations.

2.5 Summary

All of the above should be familiar to the reader, apart, maybe, from the tabular representations of the products and the artificial use of the inner and outer sets. The next step is to translate these ideas to vector products, which we do in the following chapter.

3 Vector Products

3.1 Introduction

Chapter 2 reveals some interesting patterns when we expand real, complex and quaternion products. In this chapter we review the humble scalar and cross products, and how they pave the way towards understanding geometric algebra's inner, outer and geometric products. The strategy is to understand the underlying algebra supporting all vector products, before exploring their geometric interpretation. We begin by examining the products in terms of their axioms and definitions, and then demonstrate how they are all closely related.

3.2 The scalar product

The scalar product employs a "$\cdot$" to represent the binary operator – hence its alias the *dot* product. The traditional vector analysis definition of the scalar product is:

$$a \cdot b = |a||b| \cos \varphi \tag{3.1}$$

and when we expand the product

$$a \cdot b = (a_1 e_1 + a_2 e_2 + a_3 e_3) \cdot (b_1 e_1 + b_2 e_2 + b_3 e_3) \tag{3.2}$$

we obtain

$$
\begin{aligned}
a \cdot b = \ & a_1 b_1 e_1 \cdot e_1 + a_1 b_2 e_1 \cdot e_2 + a_1 b_3 e_1 \cdot e_3 \\
& + a_2 b_1 e_2 \cdot e_1 + a_2 b_2 e_2 \cdot e_2 + a_2 b_3 e_2 \cdot e_3 \\
& + a_3 b_1 e_3 \cdot e_1 + a_3 b_2 e_3 \cdot e_2 + a_3 b_3 e_3 \cdot e_3
\end{aligned}
\tag{3.3}
$$

and invoking rule (3.1) we have

$$a \cdot b = a_1 b_1 + a_2 b_2 + a_3 b_3 \tag{3.4}$$

which is a real quantity.

J. Vince, *Geometric Algebra: An Algebraic System for Computer Games and Animation*,
© Springer-Verlag London Limited 2009

Table 3.1 shows the scalar product terms (3.4) expressed in tabular form.

TABLE 3.1   The scalar product $a \cdot b$

$a = a_1e_1 + a_2e_2 + a_3e_3$	$b = b_1e_1 + b_2e_2 + b_3e_3$		
$a \cdot b$	b_1e_1	b_2e_2	b_3e_3
a_1e_1	a_1b_1		
a_2e_2		a_2b_2	
a_3e_3			a_3b_3

Figure 3.1 gives a graphical interpretation of the scalar product, and it should be obvious that it is immaterial whether vector a is projected onto b, or vice versa.

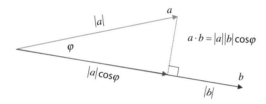

FIGURE 3.1.

Reversing the order of the vectors to

$$b \cdot a = |b||a| \cos \varphi \tag{3.5}$$

has no impact upon the result, as it only reverses the order of the scalars $|a|$ and $|b|$.

3.3 The vector product

The traditional vector analysis definition of the vector product is:

$$a \times b = c \tag{3.6}$$

where

$$|c| = |a||b| \sin \varphi \tag{3.7}$$

and φ is the positive angle between a and b.

The vector c is perpendicular to a and b, and using the right-hand rule:

$$e_1 \times e_2 = e_3 \qquad e_2 \times e_3 = e_1 \qquad e_3 \times e_1 = e_2 \tag{3.8}$$

and

$$e_i \times e_j = -e_j \times e_i. \tag{3.9}$$

Thus, expanding the vector product

$$a \times b = (a_1 e_1 + a_2 e_2 + a_3 e_3) \times (b_1 e_1 + b_2 e_2 + b_3 e_3) \tag{3.10}$$

we obtain

$$
\begin{aligned}
a \times b = &\, a_1 b_1 e_1 \times e_1 + a_1 b_2 e_1 \times e_2 + a_1 b_3 e_1 \times e_3 \\
&+ a_2 b_1 e_2 \times e_1 + a_2 b_2 e_2 \times e_2 + a_2 b_3 e_2 \times e_3 \\
&+ a_3 b_1 e_3 \times e_1 + a_3 b_2 e_3 \times e_2 + a_3 b_3 e_3 \times e_3.
\end{aligned}
\tag{3.11}
$$

Applying the rules (3.7) to (3.9) we have

$$a \times b = (a_2 b_3 - a_3 b_2) e_1 + (a_3 b_1 - a_1 b_3) e_2 + (a_1 b_2 - a_2 b_1) e_3. \tag{3.12}$$

Table 3.2 shows the vector product terms (3.12) expressed in tabular form.

TABLE 3.2   The vector product $a \times b$

$a = a_1 e_1 + a_2 e_2 + a_3 e_3$	$b = b_1 e_1 + b_2 e_2 + b_3 e_3$		
$a \times b$	$b_1 e_1$	$b_2 e_2$	$b_3 e_3$
$a_1 e_1$		$a_1 b_2 e_3$	$-a_1 b_3 e_2$
$a_2 e_2$	$-a_2 b_1 e_3$		$a_2 b_3 e_1$
$a_3 e_3$	$a_3 b_1 e_2$	$-a_3 b_2 e_1$	

Figure 3.2 shows the vector orientation for the vector product.

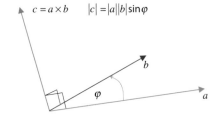

FIGURE 3.2.

Reversing the order of the vectors gives

$$b \times a = (b_1 e_1 + b_2 e_2 + b_3 e_3) \times (a_1 e_1 + a_2 e_2 + a_3 e_3) \tag{3.13}$$

which expands to

$$
\begin{aligned}
b \times a = &\, a_1 b_1 e_1 \times e_1 + a_2 b_1 e_1 \times e_2 + a_3 b_1 e_1 \times e_3 \\
&+ a_1 b_2 e_2 \times e_1 + a_2 b_2 e_2 \times e_2 + a_3 b_2 e_2 \times e_3 \\
&+ a_1 b_3 e_3 \times e_1 + a_2 b_3 e_3 \times e_2 + a_3 b_3 e_3 \times e_3
\end{aligned}
\tag{3.14}
$$

and simplifies to

$$b \times a = -(a_2 b_3 - a_3 b_2)e_1 - (a_3 b_1 - a_1 b_3)e_2 - (a_1 b_2 - a_2 b_1)e_3 \qquad (3.15)$$

and confirms that

$$a \times b = -b \times a. \qquad (3.16)$$

Table 3.3 shows the reversed vector product terms (3.15) expressed in tabular form.

TABLE 3.3   The vector product $b \times a$

$b = b_1 e_1 + b_2 e_2 + b_3 e_3 \quad a = a_1 e_1 + a_2 e_2 + a_3 e_3$			
$b \times a$	$a_1 e_1$	$a_2 e_2$	$a_3 e_3$
$b_1 e_1$		$a_2 b_1 e_3$	$-a_3 b_1 e_2$
$b_2 e_2$	$-a_1 b_2 e_3$		$a_3 b_2 e_1$
$b_3 e_3$	$a_1 b_3 e_2$	$-a_2 b_3 e_1$	

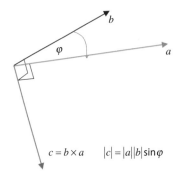

FIGURE 3.3.

Figure 3.3 shows the vector orientation for the reversed vector product.

The reader is probably familiar with this review of the scalar and vector products; nevertheless, the above description will help us understand an alternate interpretation of these products, which we will now investigate.

3.4 Dyadics

In 1884, inspired by Grassmann's theory of extensions, which identified three products of vectors: the *inner*, *exterior* and the *indeterminate* product, Josiah Willard Gibbs [1839–1903] created what he called *dyadics*. These are simple algebraic expansions of vector products consisting of *dyads*, which are indeterminate products. In fact, they were the inspiration for expanding the real, imaginary and quaternion products in the previous chapter, and in this chapter they provide the key to understanding geometric algebra.

To illustrate the relationships between classic vector analysis and geometric algebra, each product is explored algebraically and in a tabular form to identify the algebraic patterns that arise. Tables are particularly useful in this context as they reveal all the elements associated with a product, and organise them with useful visual patterns.

In an ideal world, vector products should be independent of spatial dimension – what applies to a 2D vector should also apply to a nD vector. However, this is not the case, and we have to accept possible exceptions. The real protagonist is the vector product – or its alias the *cross* product. Although this product works well in 3D it has no place in 2D. In 4D and beyond it is ambiguous, because the right-hand rule, which is central to its operation, has no automatic interpretation in spaces with more than three dimensions. However, to keep the following descriptions simple, we work with 2D and 3D vectors with the assumption that there are no restrictions in other spaces, apart from the odd one out – the cross product!

Before we can interpret a product we must declare the axioms that apply when we manipulate the individual scalar and vector terms, and as we are considering the expansion of a simple product rather than an entire algebra, we will keep these axioms to a minimum.

Knowing that commutativity is important when manipulating vectors we pay special attention to their order in the following algebra. The first two axioms describe how vectors will be allowed to interact with one another:

Axiom 1: The associative rule

$$a(bc) = (ab)c. \tag{3.17}$$

Axiom 2: The left and right distributive rules

$$a(b + c) = ab + ac$$

$$(b + c)a = ba + ca. \tag{3.18}$$

The next four axioms describe how vectors interact with scalars:
Axiom 3:
$$(\lambda a)b = \lambda(ab) = \lambda ab \qquad [\lambda \in \mathbb{R}]. \tag{3.19}$$

Axiom 4:
$$\lambda(\varepsilon a) = (\lambda \varepsilon)a \qquad [\lambda, \varepsilon \in \mathbb{R}]. \tag{3.20}$$

Axiom 5:
$$\lambda(a + b) = \lambda a + \lambda b \qquad [\lambda \in \mathbb{R}]. \tag{3.21}$$

Axiom 6:
$$(\lambda + \varepsilon)a = \lambda a + \varepsilon a \qquad [\lambda, \varepsilon \in \mathbb{R}]. \tag{3.22}$$

For example, given two vectors

$$a = a_1 e_1 + a_2 e_2 + a_3 e_3$$

$$b = b_1 e_1 + b_2 e_2 + b_3 e_3 \tag{3.23}$$

using the above axioms, their dyadic is the algebraic expansion of

$$ab = (a_1e_1 + a_2e_2 + a_3e_3)(b_1e_1 + b_2e_2 + b_3e_3) \tag{3.24}$$

which is

$$
\begin{aligned}
ab = {} & a_1b_1e_1e_1 + a_1b_2e_1e_2 + a_1b_3e_1e_3 \\
& + a_2b_1e_2e_1 + a_2b_2e_2e_2 + a_2b_3e_2e_3 \\
& + a_3b_1e_3e_1 + a_3b_2e_3e_2 + a_3b_3e_3e_3.
\end{aligned} \tag{3.25}
$$

Table 3.4 shows all the terms of the dyad ab.

TABLE 3.4   The dyad ab

$a = a_1e_1 + a_2e_2 + a_3e_3$	$b = b_1e_1 + b_2e_2 + b_3e_3$		
ab	b_1e_1	b_2e_2	b_3e_3
a_1e_1	$a_1b_1e_1e_1$	$a_1b_2e_1e_2$	$a_1b_3e_1e_3$
a_2e_2	$a_2b_1e_2e_1$	$a_2b_2e_2e_2$	$a_2b_3e_2e_3$
a_3e_3	$a_3b_1e_3e_1$	$a_3b_2e_3e_2$	$a_3b_3e_3e_3$

The dyad ab in Table 3.4 comprises two distinct sets of terms in:

$$a_1b_1e_1e_1 + a_2b_2e_2e_2 + a_3b_3e_3e_3 \tag{3.26}$$

shown in green, and

$$a_1b_2e_1e_2 + a_1b_3e_1e_3 + a_2b_1e_2e_1 + a_2b_3e_2e_3 + a_3b_1e_3e_1 + a_3b_2e_3e_2 \tag{3.27}$$

shown in pink.

Although they do not appear to represent anything obvious, the first set exhibits a greater degree of symmetry than the second. However, if we introduce some extra axioms:

Axiom 7:

$$a^2 = |a|^2 \tag{3.28}$$

Axiom 8:

$$e_1e_2 = e_3 \qquad e_2e_3 = e_1 \qquad e_3e_1 = e_2 \tag{3.29}$$

Axiom 9:

$$e_ie_j = -e_je_i \tag{3.30}$$

something interesting happens.

For example, applying these axioms to (3.25) we obtain

$$ab = a_1b_1 + a_1b_2e_3 - a_1b_3e_2$$
$$- a_2b_1e_3 + a_2b_2 + a_2b_3e_1$$
$$+ a_3b_1e_2 - a_3b_2e_1 + a_3b_3. \tag{3.31}$$

The terms of (3.31) shown in Table 3.5.

TABLE 3.5   The dyad *ab* using axioms 1 to 9

$a = a_1e_1 + a_2e_2 + a_3e_3$	$b = b_1e_1 + b_2e_2 + b_3e_3$	
ab	b_1e_1	b_2e_2	b_3e_3
a_1e_1	a_1b_1	$a_1b_2e_3$	$-a_1b_3e_2$
a_2e_2	$-a_2b_1e_3$	a_2b_2	$a_2b_3e_1$
a_3e_3	$a_3b_1e_2$	$-a_3b_2e_1$	a_3b_3

Collecting up like terms in (3.31) we have

$$ab = a_1b_1 + a_2b_2 + a_3b_3$$
$$+ (a_2b_3 - a_3b_2)e_1 + (a_3b_1 - a_1b_3)e_2 + (a_1b_2 - a_2b_1)e_3. \tag{3.32}$$

We recognise the inner set of terms as the scalar product:

$$a \cdot b = a_1b_1 + a_2b_2 + a_3b_3 \tag{3.33}$$

and the outer set of terms as the vector product:

$$a \times b = (a_2b_3 - a_3b_2)e_1 + (a_3b_1 - a_1b_3)e_2 + (a_1b_2 - a_2b_1)e_3. \tag{3.34}$$

Reversing the order of the vectors introduces a subtle change:

$$ba = (b_1e_1 + b_2e_2 + b_3e_3)(a_1e_1 + a_2e_2 + a_3e_3) \tag{3.35}$$

which expands to

$$ba = a_1b_1e_1e_1 + a_2b_1e_1e_2 + a_3b_1e_1e_3$$
$$+ a_1b_2e_2e_1 + a_2b_2e_2e_2 + a_3b_2e_2e_3$$
$$+ a_1b_3e_3e_1 + a_2b_3e_3e_2 + a_3b_3e_3e_3. \tag{3.36}$$

Applying axioms 7, 8 and 9 to (3.36) we obtain

$$ba = a_1b_1 + a_2b_1e_3 - a_3b_1e_2$$
$$- a_1b_2e_3 + a_2b_2 + a_3b_2e_1$$
$$+ a_1b_3e_2 - a_2b_3e_1 + a_3b_3 \tag{3.37}$$

which simplifies to

$$ba = a_1b_1 + a_2b_2 + a_3b_3$$
$$- [(a_2b_3 - a_3b_2)e_1 + (a_3b_1 - a_1b_3)e_2 + (a_1b_2 - a_2b_1)e_3] \,. \tag{3.38}$$

The two sets of terms still remain but the sign of the vector part is reversed. Thus we observe that using the above axioms, the scalar part is symmetric, and the vector part is antisymmetric. In fact, using the above axioms, the dyad ab is the sum of the scalar product and the vector product:

$$ab = a \cdot b + a \times b. \tag{3.39}$$

At this point we are getting very close to the foundations of geometric algebra. So let's take a closer look at these two products.

3.5 The outer product

3.5.1 Origins of the outer product

When the equivalent of the expression

$$a_1b_2e_1e_2 + a_1b_3e_1e_3 + a_2b_1e_2e_1 + a_2b_3e_2e_3 + a_3b_1e_3e_1 + a_3b_2e_3e_2 \tag{3.40}$$

appeared in Hamilton's original quaternion product, it was assumed that it described a vector, which is one reason why vector analysis started life using the scalar and vector products. However, the mathematician, Hermann Grassmann [1809–1877], proposed an alternative *exterior* product, but his voice was not heard until much later when its true significance was appreciated. By which time, the direction of 3D vector analysis had been determined by Josiah Gibbs and Hamilton's, as well as Grassmann's, ideas were discarded.

Unlike the vector product, Grassmann's *exterior* product works in 2D, 3D, . . . nD. The operator used to denote the product is the wedge symbol "∧" – hence its alias the *wedge* product, and is written $a \wedge b$. Therefore, given two vectors

$$a = a_1e_1 + a_2e_2 + a_3e_3$$
$$b = b_1e_1 + b_2e_2 + b_3e_3 \tag{3.41}$$

their exterior product is

$$a \wedge b = (a_1e_1 + a_2e_2 + a_3e_3) \wedge (b_1e_1 + b_2e_2 + b_3e_3) \tag{3.42}$$

which expands to

$$a \wedge b = a_1b_1e_1 \wedge e_1 + a_1b_2e_1 \wedge e_2 + a_1b_3e_1 \wedge e_3$$
$$+ a_2b_1e_2 \wedge e_1 + a_2b_2e_2 \wedge e_2 + a_2b_3e_2 \wedge e_3$$
$$+ a_3b_1e_3 \wedge e_1 + a_3b_2e_3 \wedge e_2 + a_3b_3e_3 \wedge e_3. \tag{3.43}$$

Apart from the axioms stated in (3.1) to (3.6), the extra axioms associated with the exterior or what has become the outer product are:

$$e_i \wedge e_i = 0 \qquad e_i \wedge e_j = -e_j \wedge e_i \qquad (3.44)$$

which allow us to write (3.43) as

$$a \wedge b = a_1 b_2 e_1 \wedge e_2 - a_1 b_3 e_3 \wedge e_1$$
$$- a_2 b_1 e_1 \wedge e_2 - a_2 b_3 e_2 \wedge e_3$$
$$+ a_3 b_1 e_3 \wedge e_1 - a_3 b_2 e_2 \wedge e_3. \qquad (3.45)$$

Note that the wedge product subscripts have been adjusted to retain the cyclic sequence: 1,2 2,3 3,1 which introduces some useful symmetry. Table 3.6 shows the outer product terms (3.45) in tabular form.

TABLE 3.6   The outer product $a \wedge b$

$a = a_1e_1 + a_2e_2 + a_3e_3 \quad b = b_1e_1 + b_2e_2 + b_3e_3$			
$a \wedge b$	b_1e_1	b_2e_2	b_3e_3
a_1e_1		$a_1b_2e_1 \wedge e_2$	$-a_1b_3e_3 \wedge e_1$
a_2e_2	$-a_2b_1e_1 \wedge e_2$		$a_2b_3e_2 \wedge e_3$
a_3e_3	$a_3b_1e_3 \wedge e_1$	$-a_3b_2e_2 \wedge e_3$	

Collecting up like terms in (3.45) we have

$$a \wedge b = (a_2b_3 - a_3b_2)e_2 \wedge e_3 + (a_3b_1 - a_1b_3)e_3 \wedge e_1 + (a_1b_2 - a_2b_1)e_1 \wedge e_2 \qquad (3.46)$$

which looks very familiar! In fact, replacing $e_2 \wedge e_3$, $e_3 \wedge e_1$, $e_1 \wedge e_2$ by e_1, e_2, e_3 respectively we recover the vector product in (3.34). Therefore, it is obvious that the outer and vector products are intimately related.

From (3.46) we see that $a \wedge b$ is the sum of a series of terms defined by

$$(a \wedge b)_{ij} = (a_ib_j - a_jb_i)e_i \wedge e_j \qquad (i, j = 1, 2, 3). \qquad (3.47)$$

What we must now uncover is the geometric meaning of the terms $e_2 \wedge e_3$, $e_3 \wedge e_1$ and $e_1 \wedge e_2$, which, fortunately, happens to the rather simple.

3.5.2 The geometric meaning of the outer product in 2D

Although we have an algebraic definition of the outer product, it would be useful to have a geometric interpretation, so that we can visualise what is going on. To help us in this endeavour, we start with two 2D vectors a and b

$$a = a_1e_1 + a_2e_2$$
$$b = b_1e_1 + b_2e_2 \qquad (3.48)$$

whose outer product is

$$a \wedge b = (a_1 e_1 + a_2 e_2) \wedge (b_1 e_1 + b_2 e_2)$$

$$= (a_1 b_2 - a_2 b_1) e_1 \wedge e_2 \qquad (3.49)$$

which contains two terms: a scalar $a_1 b_2 - a_2 b_1$ and something new $e_1 \wedge e_2$ called a *unit bivector*.

The scalar term $a_1 b_2 - a_2 b_1$ is the area of the parallelogram formed by the vectors a and b. We can demonstrate this with reference to Figure 3.4 that shows the areas of the internal rectangles and triangles. The area of the blue parallelogram is the area of the outer rectangle less the inner areas:

$$area = (a_1 + b_1)(a_2 + b_2) - \frac{1}{2}b_1 b_2 - a_2 b_1 - \frac{1}{2}a_1 a_2 - \frac{1}{2}b_1 b_2 - a_2 b_1 - \frac{1}{2}a_1 a_2$$

$$area = a_1 b_2 - a_2 b_1. \qquad (3.50)$$

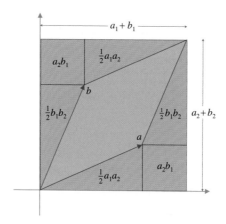

FIGURE 3.4.

The unit bivector, on the other hand, is a new mathematical object and requires explaining, but before we do so, let's review one final piece of geometry relevant to the outer product.

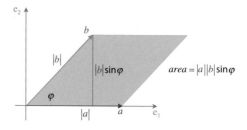

FIGURE 3.5.

Figure 3.5 shows two vectors a and b with the parallelogram associated with their addition. The second vector b is deliberately rotated through a positive angle φ relative to a. Reversing this relationship gives rise to a negative angle with a corresponding negative sine. The figure reminds us that the area of the parallelogram is $|a||b| \sin \varphi$, therefore we can state confidently that

$$|a||b| \sin \varphi = |a_1 b_2 - a_2 b_1|. \tag{3.51}$$

Reversing the vectors in the outer product produces

$$b \wedge a = (b_1 e_1 + b_2 e_2) \wedge (a_1 e_1 + a_2 e_2) \tag{3.52}$$

which, using the axioms in (3.44), expands to

$$b \wedge a = -(a_1 b_2 - a_2 b_1) e_1 \wedge e_2. \tag{3.53}$$

Observe that reversing the vector sequence reverses the sign of the product, therefore

$$a \wedge b = -b \wedge a \tag{3.54}$$

and like the vector product, the outer product of two vectors is antisymmetric.

We are now in a position to give a geometric interpretation of $a \wedge b$ and $e_1 \wedge e_2$.

We already know that the components of a vector define its orientation. Similarly, the orientation of a surface is defined by its bivector components. Thus when we encounter an expression such as

$$a \wedge b = (a_1 b_2 - a_2 b_1) e_1 \wedge e_2 \tag{3.55}$$

the scalar term $a_1 b_2 - a_2 b_1$ defines a signed area, whilst the unit bivector term $e_1 \wedge e_2$ identifies the plane associated with the area. So the bivector term allows us to interpret the orientation of the area, which requires us to define a reference orientation.

We begin with a definition for 2D space as shown in Figure 3.6, with the two unit basis vectors e_1 and e_2 where e_2 is swept along the length of e_1 in two arbitrary steps. By definition, this orientation is defined as positive, and if the outer product of a pair of vectors is also positive, it implies that the vectors are oriented in the same way as the unit basis vectors. On the other hand, if the outer product of two vectors is negative, it implies that their orientation opposes that of the basis vectors.

Thus the bivector term $e_1 \wedge e_2$ reminds us that the "sense" or orientation of the surface is anticlockwise, as depicted by the directed circle in Figure 3.6.

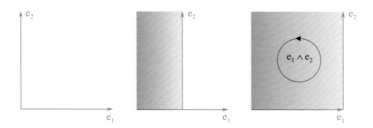

FIGURE 3.6.

Conversely, Figure 3.7 shows the basis vectors e_1 and e_2 where e_1 is swept along the length of e_2 in two arbitrary steps. This time the bivector term $e_2 \wedge e_1$ informs us that the "sense" or orientation of the surface is clockwise, as depicted by the directed circle in Figure 3.7.

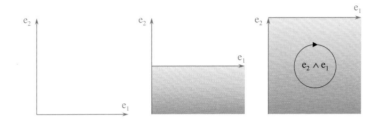

FIGURE 3.7.

Needless to say, the positive orientation of 2D space is represented by $e_1 \wedge e_2$, and is widely used throughout the mathematics community. Perhaps the best way to illustrate this description is through an example.

Given two vectors

$$a = 3e_1 + e_2$$
$$b = 2e_1 + 3e_2 \tag{3.56}$$

their outer product is

$$a \wedge b = (3e_1 + e_2) \wedge (2e_1 + 3e_2) \tag{3.57}$$

which expands to

$$a \wedge b = 6e_1 \wedge e_1 + 9e_1 \wedge e_2 + 2e_2 \wedge e_1 + 3e_2 \wedge e_2 \tag{3.58}$$

and reduces to

$$a \wedge b = 7e_1 \wedge e_2 \tag{3.59}$$

which confirms that the vectors are oriented in an anticlockwise sequence and the area of the associated parallelogram is 7.

On the other hand, the outer product $b \wedge a$ is

$$b \wedge a = (2e_1 + 3e_2) \wedge (3e_1 + e_2) \tag{3.60}$$

which expands to

$$= 6e_1 \wedge e_1 + 2e_1 \wedge e_2 + 9e_2 \wedge e_1 + 3e_2 \wedge e_2 \tag{3.61}$$

and reduces to

$$b \wedge a = 7e_2 \wedge e_1 = -7e_1 \wedge e_2 \tag{3.62}$$

which confirms that the vectors are oriented in a clockwise sequence. Thus the unit bivector tells us how the basis vectors are organised.

Now that the outer product of two vectors is defined, let's examine the meaning of a triple outer product. Starting with three 2D vectors

$$a = a_1 e_1 + a_2 e_2$$
$$b = b_1 e_1 + b_2 e_2$$
$$c = c_1 e_1 + c_2 e_2 \tag{3.63}$$

their outer product is

$$a \wedge b \wedge c = (a_1 e_1 + a_2 e_2) \wedge (b_1 e_1 + b_2 e_2) \wedge (c_1 e_1 + c_2 e_2)$$
$$= (a_1 b_2 - a_2 b_1)(e_1 \wedge e_2) \wedge (c_1 e_1 + c_2 e_2). \tag{3.64}$$

At this point in the expansion we can see from (3.64) that we have to form the outer product between $(e_1 \wedge e_2) \wedge e_1$ and $(e_1 \wedge e_2) \wedge e_2$, both of which are zero. The reason for this is that the outer product is associative (see Appendix A), which means that using the axioms of (3.44)

$$a \wedge (b \wedge c) = (a \wedge b) \wedge c = a \wedge b \wedge c \tag{3.65}$$

therefore

$$(e_1 \wedge e_2) \wedge e_1 = e_1 \wedge e_2 \wedge e_1 = -e_2 \wedge e_1 \wedge e_1 = 0 \tag{3.66}$$

and

$$(e_1 \wedge e_2) \wedge e_2 = e_1 \wedge e_2 \wedge e_2 = e_1 \wedge e_2 \wedge e_2 = 0. \tag{3.67}$$

Thus the outer product of three or more 2D vectors is zero.

3.5.3 The geometric meaning of the outer product in 3D

We have seen above that the outer product of two 2D basis vectors provides a mechanism to describe the orientation of a 2D plane. In 3D there are three unit basis vectors and consequently, three unit bivectors. Let's explore how they are computed.

To begin, consider the right-handed axial system shown in Figure 3.8a. The unit bivector describing the planar surface containing e_1 and e_2 is $e_1 \wedge e_2$ as shown in Figure 3.8b, whereas the unit bivector $e_2 \wedge e_3$ describes the planar surface containing e_2 and e_3, and the unit bivector $e_3 \wedge e_1$ describes the planar surface containing e_3 and e_1. Therefore given two 3D vectors:

$$a = a_1 e_1 + a_2 e_2 + a_3 e_3$$
$$b = b_1 e_1 + b_2 e_2 + b_3 e_3 \tag{3.68}$$

their outer product is

$$a \wedge b = (a_1 b_2 - a_2 b_1)e_1 \wedge e_2 + (a_2 b_3 - a_3 b_2)e_2 \wedge e_3 + (a_3 b_1 - a_1 b_3)e_3 \wedge e_1. \tag{3.69}$$

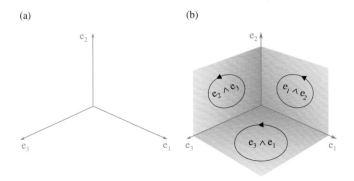

FIGURE 3.8.

The outer product $a \wedge b$ can be visualized as a parallelogram in 3-space, which makes projections upon the three unit basis bivectors with areas:

$$area_{12} = a_1 b_2 - a_2 b_1 \text{ is the proejction upon } e_1 \wedge e_2$$

$$area_{23} = a_2 b_3 - a_3 b_2 \text{ is the proejction upon } e_2 \wedge e_3$$

$$area_{31} = a_3 b_1 - a_1 b_3 \text{ is the proejction upon } e_3 \wedge e_1. \tag{3.70}$$

Figure 3.9 illustrates this relationship, and we see that

$$a_{12} \wedge b_{12} = (a_1 b_2 - a_2 b_1)e_1 \wedge e_2$$

$$a_{23} \wedge b_{23} = (a_2 b_3 - a_3 b_2)e_2 \wedge e_3$$

$$a_{31} \wedge b_{31} = (a_3 b_1 - a_1 b_3)e_3 \wedge e_1 \tag{3.71}$$

thus

$$a \wedge b = a_{12} \wedge b_{12} + a_{23} \wedge b_{23} + a_{31} \wedge b_{31}. \tag{3.72}$$

But what is the relationship between the area of the parallelogram formed by $a \wedge b$ and the three projected areas $area_{12}$, $area_{23}$ and $area_{31}$? We can answer this question with the following analysis:
We already know that

$$|a \wedge b| = |a||b| \sin \varphi \tag{3.73}$$

therefore,

$$|a \wedge b|^2 = |a|^2 |b|^2 \sin^2 \varphi$$

$$= |a|^2 |b|^2 (1 - \cos^2 \varphi). \tag{3.74}$$

From the definition of the scalar product we have

$$\cos^2 \varphi = \frac{(a_1 b_2 + a_2 b_2 + a_3 b_3)^2}{|a|^2 |b|^2} \tag{3.75}$$

therefore,

$$|a \wedge b|^2 = |a|^2|b|^2 \left(1 - \frac{(a_1b_1 + a_2b_2 + a_3b_3)^2}{|a|^2|b|^2}\right)$$

$$= |a|^2|b|^2 - (a_1b_1 + a_2b_2 + a_3b_3)^2$$

$$= (a_1^2 + a_2^2 + a_3^2)(b_1^2 + b_2^2 + b_3^2) - (a_1b_1 + a_2b_2 + a_3b_3)^2 \tag{3.76}$$

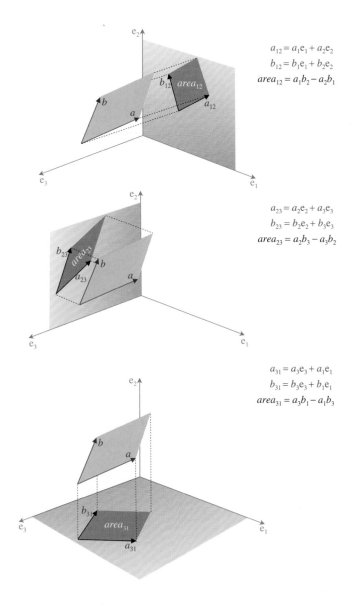

$$a_{12} = a_1\mathbf{e}_1 + a_2\mathbf{e}_2$$
$$b_{12} = b_1\mathbf{e}_1 + b_2\mathbf{e}_2$$
$$area_{12} = a_1b_2 - a_2b_1$$

$$a_{23} = a_2\mathbf{e}_2 + a_3\mathbf{e}_3$$
$$b_{23} = b_2\mathbf{e}_2 + b_3\mathbf{e}_3$$
$$area_{23} = a_2b_3 - a_3b_2$$

$$a_{31} = a_3\mathbf{e}_3 + a_1\mathbf{e}_1$$
$$b_{31} = b_3\mathbf{e}_3 + b_1\mathbf{e}_1$$
$$area_{31} = a_3b_1 - a_1b_3$$

FIGURE 3.9.

which simplifies to

$$= (a_1^2 b_2^2 - 2a_1 a_2 b_1 b_2 + a_2^2 b_1^2) + (a_2^2 b_3^2 - 2a_2 a_3 b_2 b_3 + a_3^2 b_2^2)$$
$$+ (a_3^2 b_1^2 - 2a_3 a_1 b_3 b_1 + a_1^2 b_3^2)$$

$$|a \wedge b|^2 = (a_1 b_2 - a_2 b_1)^2 + (a_2 b_3 - a_3 b_2)^2 + (a_3 b_1 - a_1 b_3)^2 \tag{3.77}$$

therefore,

$$|a \wedge b|^2 = area_{12}^2 + area_{23}^2 + area_{31}^2. \tag{3.78}$$

Thus the square of a bivector's area equals the sum of the squares of the areas projected onto the basis unit bivectors.

Let's test (3.78) with an example.

Given two vectors

$$a = 2e_1 + 3e_2 + 5e_3$$
$$b = 2e_1 + 4e_2 + 7e_3 \tag{3.79}$$

then

$$area_{12} = (2 \times 4 - 2 \times 3) = 2$$
$$area_{23} = (3 \times 7 - 4 \times 5) = 1$$
$$area_{31} = (5 \times 2 - 7 \times 2) = -4 \tag{3.80}$$

therefore,

$$|a \wedge b|^2 = 2^2 + 1^2 + (-4)^2 = 21 \tag{3.81}$$

and the area of the parallelogram formed by $a \wedge b$ is $\sqrt{21}$.

Bearing in mind that the outer product is intimately related to the vector product, this relationship should not be too much of a surprise. The three areas are the vector components of the axial vector formed by the vector product, whose length is related to the components via the Pythagorean rule.

The product of two vectors in 3D results in the projection of areas onto the three unit basis bivectors, which collectively encode the geometric characteristics of the original vectors. Next is the outer product of three vectors, which is where we discover the natural progression from a moving point to create a line, which when moved along another line creates an area, which in turn when moved along another line creates a volume.

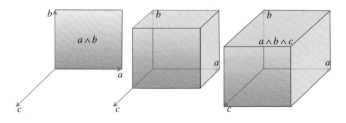

Figure 3.10.

Consider the scenario shown in Figure 3.10 where we see a bivector $a \wedge b$ swept along a third vector c in two arbitrary steps. Assuming that the vectors are mutually orthogonal, we have no problem in accepting that the volume is $|a||b||c|$. If we now assume that the vectors are not orthogonal, as shown in Figure 3.11, the area of the parallelogram formed by a and b is

$$area = |a||b| \sin \theta \tag{3.82}$$

where θ is the angle between the vectors, which we know is $a \wedge b$. If this area is now swept along c, which makes an angle ϕ with a and b, a volume is created, whose value is

$$volume = |a||b| \sin \theta |c| \sin \phi \tag{3.83}$$

which is nothing more than the outer product of the three vectors

$$volume = a \wedge b \wedge c. \tag{3.84}$$

From our knowledge of volumes, we know that a volume is independent of how the sides are multiplied together, which means that if we maintain the relative order between the vectors:

$$a \wedge b \wedge c = b \wedge c \wedge a = c \wedge a \wedge b. \tag{3.85}$$

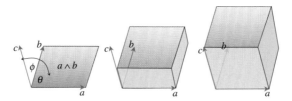

FIGURE 3.11.

Before demonstrating that (3.84) holds for any three vectors, let's explore what happens when three unit basis vectors are combined using the outer product.

There are only three possible combinations for the unit basis vectors:

- all three unit basis vectors are identical
- two unit basis vectors are identical
- all unit basis vectors are different.

An example of the first combination is $e_1 \wedge e_1 \wedge e_1$ which must be zero because $e_1 \wedge e_1 = 0$.

An example of the second combination is $e_1 \wedge e_2 \wedge e_1$ which is also zero because

$$e_1 \wedge e_2 \wedge e_1 = -(e_1 \wedge e_1) \wedge e_2 = 0. \tag{3.86}$$

An example of the third combination is $e_1 \wedge e_2 \wedge e_3$ whose magnitude must be 1. The reason being that each basis vector has a magnitude of 1, and they are mutually orthogonal. In fact, this is a totally new mathematical object called a *trivector* and is associated with the orientation of a volume, in exactly the same way a bivector is associated with the orientation of a planar surface. The convention is that if the vectors are combined as shown in Figure 3.10 or 3.11, a positive volume results, otherwise it is negative. For example, if we swapped the original vectors to $b \wedge a$, the starting area is negative, which when swept along vector c results in a negative volume:

$$b \wedge a \wedge c = -a \wedge b \wedge c. \tag{3.87}$$

Although we can expand the outer product of three vectors algebraically, it does become rather fussy. An alternative is to use a table, which is compact, but in this case is a 3D table. Therefore, starting with three vectors

$$a = a_1 e_1 + a_2 e_2 + a_3 e_3$$
$$b = b_1 e_1 + b_2 e_2 + b_3 e_3$$
$$c = c_1 e_1 + c_2 e_2 + c_3 e_3 \tag{3.88}$$

their complete algebraic product abc is shown in Table 3.7 using 3 dimensions. The vertical direction is a, the horizontal direction is b, whilst the depth direction is c.

There are 27 entries in all. A quick inspection of the table reveals that many of the entries collapse to zero when we apply the rules of the outer product. This means that any entry containing two or three identical basis vectors disappears. In fact, only 6 terms remain, as shown in Table 3.8. Note

TABLE 3.7   The expansion of the triad abc

$c_3 e_3$	$b_1 e_1$	$b_2 e_2$	$b_3 e_3$
$a_1 e_1$	$a_1 b_1 c_3$ $e_1 e_1 e_3$	$a_1 b_2 c_3$ $e_1 e_2 e_3$	$a_1 b_3 c_3$ $e_1 e_3 e_3$
$a_2 e_2$	$a_2 b_1 c_3$ $e_2 e_1 e_3$	$a_2 b_2 c_3$ $e_2 e_2 e_3$	$a_2 b_3 c_3$ $e_2 e_3 e_3$
$a_3 e_3$	$a_3 b_1 c_3$ $e_3 e_1 e_3$	$a_3 b_2 c_3$ $e_3 e_2 e_3$	$a_3 b_3 c_3$ $e_3 e_3 e_3$

$c_2 e_2$	$b_1 e_1$	$b_2 e_2$	$b_3 e_3$
$a_1 e_1$	$a_1 b_1 c_2$ $e_1 e_1 e_2$	$a_1 b_2 c_2$ $e_1 e_2 e_2$	$a_1 b_3 c_2$ $e_1 e_3 e_2$
$a_2 e_2$	$a_2 b_1 c_2$ $e_2 e_1 e_2$	$a_2 b_2 c_2$ $e_2 e_2 e_2$	$a_2 b_3 c_2$ $e_2 e_3 e_2$
$a_3 e_3$	$a_3 b_1 c_2$ $e_3 e_1 e_2$	$a_3 b_2 c_2$ $e_3 e_2 e_2$	$a_3 b_3 c_2$ $e_3 e_3 e_2$

$c_1 e_1$	$b_1 e_1$	$b_2 e_2$	$b_3 e_3$
$a_1 e_1$	$a_1 b_1 c_1$ $e_1 e_1 e_1$	$a_1 b_2 c_1$ $e_1 e_2 e_1$	$a_1 b_3 c_1$ $e_1 e_3 e_1$
$a_2 e_2$	$a_2 b_1 c_1$ $e_2 e_1 e_1$	$a_2 b_2 c_1$ $e_2 e_2 e_1$	$a_2 b_3 c_1$ $e_2 e_3 e_1$
$a_3 e_3$	$a_3 b_1 c_1$ $e_3 e_1 e_1$	$a_3 b_2 c_1$ $e_3 e_2 e_1$	$a_3 b_3 c_1$ $e_3 e_3 e_1$

TABLE 3.8   The outer product $a \wedge b \wedge c$

	c_3e_3	b_1e_1	b_2e_2	b_3e_3
a_1e_1			$a_1b_2c_3$ $e_1 \wedge e_2 \wedge e_3$	
a_2e_2		$-a_2b_1c_3$ $e_1 \wedge e_2 \wedge e_3$		
a_3e_3				

	c_2e_2	b_1e_1	b_2e_2	b_3e_3
a_1e_1				$-a_1b_3c_2$ $e_1 \wedge e_2 \wedge e_3$
a_2e_2				
a_3e_3		$a_3b_1c_2$ $e_1 \wedge e_2 \wedge e_3$		

	c_1e_1	b_1e_1	b_2e_2	b_3e_3
a_1e_1				
a_2e_2				$a_2b_3c_1$ $e_1 \wedge e_2 \wedge e_3$
a_3e_3			$-a_3b_2c_1$ $e_1 \wedge e_2 \wedge e_3$	

that the unit basis trivectors have been rearranged to create a consistent basis vector sequence $e_1e_2e_3$, which switches the sign for some terms.

From Table 3.8 we see that

$$a \wedge b \wedge c = a_1b_2c_3 e_1 \wedge e_2 \wedge e_3 + a_2b_3c_1 e_1 \wedge e_2 \wedge e_3 + a_3b_1c_2 e_1 \wedge e_2 \wedge e_3$$
$$- a_1b_3c_2 e_1 \wedge e_2 \wedge e_3 - a_2b_1c_3 e_1 \wedge e_2 \wedge e_3 - a_3b_2c_1 e_1 \wedge e_2 \wedge e_3 \qquad (3.89)$$

or

$$a \wedge b \wedge c = (a_1b_2c_3 + a_2b_3c_1 + a_3b_1c_2 - a_1b_3c_2 - a_2b_1c_3 - a_3b_2c_1)e_1 \wedge e_2 \wedge e_3 \qquad (3.90)$$

or as a determinant

$$a \wedge b \wedge c = \begin{vmatrix} a_1 & b_1 & c_1 \\ a_2 & b_2 & c_2 \\ a_3 & b_3 & c_3 \end{vmatrix} e_1 \wedge e_2 \wedge e_3. \qquad (3.91)$$

The determinant in (3.91) is the signed volume of a parallelpiped, whilst the unit basis trivector term reminds us of the orientation.

3.6 Summary

When Hamilton invented quaternions, he simultaneously invented vectors, which proved to be an invaluable mathematical object for representing quantities possessing magnitude and direction. In the current context the word "orientation" is preferred to "direction". Within a relatively short period of time, mathematicians and physicists of the day decided to abandon quaternions and adopt vectors and two products: the scalar and vector products. The scalar product results in a scalar, whereas the vector product creates a third vector perpendicular to the original vectors.

At the time no one realized the negative effect this would have upon the progress of vector analysis, apart from Grassmann, who had put forward his inner and outer products as an alternative to the scalar and vector products. Grassmann's outer product takes two vectors and creates a new mathematical object: a bivector, which is ideal for representing areas with magnitude and orientation. The product of three independent vectors creates a trivector, which is another mathematical object that represents volumes with magnitude and orientation. The advantage of the outer product is that it continues to work in spaces of any dimension, whereas the vector product is really only at home in three dimensions.

As we have seen above, there is a very close relationship between the vector and outer product, and later on we will see that it is very simple to convert one into the other. In the next chapter we discover how a third product opens up the true world of geometric algebra.

4 The Geometric Product

4.1 Introduction

In this chapter we introduce the geometric product which unites the inner (scalar) product with the outer product. The resulting algebra opens up a totally new way for manipulating vectors and their close relations bivectors and trivectors and higher-order objects.

4.2 Axioms

Before Josiah Willard Gibbs had thought of dyadics William Kingdon Clifford [1845–1879] realised that it was possible to unite the scalar and outer products into a single product called the *geometric* product – it was just a question of adopting an appropriate set of axioms, which we have already encountered. One of the axioms, the modulus axiom (3.28), can also be defined as

$$a^2 = \pm |a|^2 \tag{4.1}$$

which allows the square of a vector to be negative as well as positive. This strange axiom is useful when applying geometric algebra to non-Euclidean spaces. But for the moment we will keep to the original definition:

$$a^2 = + |a|^2 \tag{4.2}$$

which ensures that

$$e_i^2 = 1. \tag{4.3}$$

At this point it is worth introducing some shorthand notation when manipulating basis vector products:

$$e_i e_j \cdots e_n \equiv e_{ij\cdots n}. \tag{4.4}$$

J. Vince, *Geometric Algebra: An Algebraic System for Computer Games and Animation*,
© Springer-Verlag London Limited 2009

For example,

$$e_1 e_2 e_3 \equiv e_{123}. \tag{4.5}$$

Furthermore, the modulus axiom allows us to make the following reduction:

$$e_i e_i e_j = e_i^2 e_j = e_j. \tag{4.6}$$

For example,

$$e_{21} = -e_{12}$$
$$e_{312} = e_{123}$$
$$e_{112} = e_2$$
$$e_{121} = -e_2. \tag{4.7}$$

The geometric product of two vectors is defined as

$$ab = a \cdot b + a \wedge b \tag{4.8}$$

which effectively divides the dyad ab into two parts - the inner and outer products.

To understand how the geometric product behaves, we apply it to different vector combinations and begin with two identical basis vectors:

$$e_i e_i = e_i \cdot e_i + e_i \wedge e_i = e_i^2 = 1 \tag{4.9}$$

and the product of two orthogonal basis vectors is

$$e_i e_j = e_i \cdot e_j + e_i \wedge e_j = e_i \wedge e_j. \tag{4.10}$$

Consequently, the geometric product of two 2D vectors

$$a = a_1 e_1 + a_2 e_2$$
$$b = b_1 e_1 + b_2 e_2$$

is

$$
\begin{aligned}
ab &= (a_1 e_1 + a_2 e_2) \cdot (b_1 e_1 + b_2 e_2) + (a_1 e_1 + a_2 e_2) \wedge (b_1 e_1 + b_2 e_2) \\
&= (a_1 b_1 e_1 \cdot e_1 + a_2 b_2 e_2 \cdot e_2) + (a_1 b_2 e_1 \wedge e_2 + a_2 b_1 e_2 \wedge e_1) \\
&= (a_1 b_1 + a_2 b_2) + (a_1 b_2 - a_2 b_1) e_1 \wedge e_2 \\
&= (a_1 b_1 + a_2 b_2) + (a_1 b_2 - a_2 b_1) e_{12}.
\end{aligned} \tag{4.11}
$$

Note how e_{12} has been substituted for $e_1 \wedge e_2$. This is not only algebraically correct, but is much more compact. Table 4.1 shows the geometric product terms (4.11) in tabular form.

TABLE 4.1   The geometric product ab

$a = a_1e_1 + a_2e_2 \quad b = b_1e_1 + b_2e_2$		
ab	b_1e_1	b_2e_2
a_1e_1	a_1b_1	$a_1b_2e_{12}$
a_2e_2	$-a_2b_1e_{12}$	a_2b_2

The reverse geometric product ba is

$$ba = (b_1e_1 + b_2e_2) \cdot (a_1e_1 + a_2e_2) + (b_1e_1 + b_2e_2) \wedge (a_1e_1 + a_2e_2)$$
$$= (a_1b_1e_1 \cdot e_1 + a_2b_2e_2 \cdot e_2) + (a_2b_1e_1 \wedge e_2 + a_1b_2e_2 \wedge e_1)$$
$$= (a_1b_1 + a_2b_2) - (a_1b_2 - a_2b_1)e_1 \wedge e_2$$
$$= (a_1b_1 + a_2b_2) - (a_1b_2 - a_2b_1)e_{12}. \tag{4.12}$$

Table 4.2 shows the geometric product terms (4.12) in tabular form.

TABLE 4.2   The geometric product ba

$b = b_1e_1 + b_2e_2 \quad a = a_1e_1 + a_2e_2$		
ba	a_1e_1	a_2e_2
b_1e_1	a_1b_1	$a_2b_1e_{12}$
b_2e_2	$-a_1b_2e_{12}$	a_2b_2

Note that the signs of the outer product terms have been swapped.

So the geometric product of two 2D vectors results in the sum of a scalar term and a bivector, which makes an unlikely combination. But let's not be unnerved by this weird duo, for if the algebra gives rise to such combinations, then there is probably a good reason for their existence.

Now let's examine the product of two 3D vectors

$$a = a_1e_1 + a_2e_2 + a_3e_3$$
$$b = b_1e_1 + b_2e_2 + b_3e_3. \tag{4.13}$$

Their geometric product is

$$ab = (a_1e_1 + a_2e_2 + a_3e_3) \cdot (b_1e_1 + b_2e_2 + b_3e_3)$$
$$+ (a_1e_1 + a_2e_2 + a_3e_3) \wedge (b_1e_1 + b_2e_2 + b_3e_3)$$

$$= a_1 b_1 + a_2 b_2 + a_3 b_3$$
$$+ (a_1 b_2 - a_2 b_1) e_1 \wedge e_2$$
$$+ (a_2 b_3 - a_3 b_2) e_2 \wedge e_3$$
$$+ (a_3 b_1 - a_1 b_3) e_3 \wedge e_1$$

$$= a_1 b_1 + a_2 b_2 + a_3 b_3$$
$$+ (a_1 b_2 - a_2 b_1) e_{12} + (a_2 b_3 - a_3 b_2) e_{23} + (a_3 b_1 - a_1 b_3) e_{31}. \qquad (4.14)$$

Table 4.3 shows the geometric product terms (4.14) in tabular form.

TABLE 4.3   The geometric product ab

$a = a_1 e_1 + a_2 e_2 + a_3 e_3$	$b = b_1 e_1 + b_2 e_2 + b_3 e_3$		
ab	$b_1 e_1$	$b_2 e_2$	$b_3 e_3$
$a_1 e_1$	$a_1 b_1$	$a_1 b_2 e_{12}$	$-a_1 b_3 e_{31}$
$a_2 e_2$	$-a_2 b_1 e_{12}$	$a_2 b_2$	$a_2 b_3 e_{23}$
$a_3 e_3$	$a_3 b_1 e_{31}$	$-a_3 b_2 e_{23}$	$a_3 b_3$

The geometric product, then, is nothing more than the dyad ab interpreted using a slightly different set of axioms.

The reverse geometric product ba is

$$ba = (b_1 e_1 + b_2 e_2 + b_3 e_3) \cdot (a_1 e_1 + a_2 e_2 + a_3 e_3)$$
$$+ (b_1 e_1 + b_2 e_2 + b_3 e_3) \wedge (a_1 e_1 + a_2 e_2 + a_3 e_3)$$

$$= a_1 b_1 + a_2 b_2 + a_3 b_3$$
$$- (a_1 b_2 - a_2 b_1) e_1 \wedge e_2$$
$$- (a_2 b_3 - a_3 b_2) e_2 \wedge e_3$$
$$- (a_3 b_1 - a_1 b_3) e_3 \wedge e_1$$

$$= a_1 b_1 + a_2 b_2 + a_3 b_3$$
$$- (a_1 b_2 - a_2 b_1) e_{12} - (a_2 b_3 - a_3 b_2) e_{23} - (a_3 b_1 - a_1 b_3) e_{31}. \qquad (4.15)$$

Table 4.4 shows the geometric product terms (4.15) in tabular form.

Note again how the sign of the outer product terms have been swapped.

TABLE 4.4   The geometric product ba

$b = b_1e_1 + b_2e_2 + b_3e_3 \quad a = a_1e_1 + a_2e_2 + a_3e_3$			
ba	a_1e_1	a_2e_2	a_3e_3
b_1e_1	a_1b_1	$a_2b_1e_{12}$	$-a_3b_1e_{31}$
b_2e_2	$-a_1b_2e_{12}$	a_2b_2	$a_3b_2e_{23}$
b_3e_3	$a_1b_3e_{31}$	$-a_2b_3e_{23}$	a_3b_3

This time the geometric product of two 3D vectors results in the sum of a scalar term and three bivector terms, and geometric algebra is beginning to reveal some of its secrets.

The names "inner" and "outer" product were proposed by Grassmann, while Hamilton proposed the names "scalar" and "vector" product. However, there is a subtle difference between Grassmann's outer product and Hamilton's vector product. The outer product computes signed areas for each basis bivector, whereas the vector product creates an axial vector perpendicular to the original vectors. Furthermore, the outer product functions in any finite dimension (apart from 1D), whereas the vector product is unambiguous in dimensions other than in 3D.

At this point we abandon the name *scalar* product for the alternative *inner* product. The main reason for this substitution is that the geometric product not only multiplies vectors together but any combination of vectors, bivectors and trivectors. In these situations the result is not necessarily a scalar, and the scalar prefix is not appropriate. Thus the geometric product unites the inner and outer products in a similar way that the quaternion product unites the scalar and vector products.

4.3 Redefining the inner and outer products

As the geometric product is defined in terms of the inner and outer products (4.8) it seems only natural to expect that similar functions exist relating the inner and outer products in terms of the geometric product. Such functions do exist and emerge when we combine the following two equations:

$$ab = a \cdot b + a \wedge b \tag{4.16}$$

$$ba = a \cdot b - a \wedge b. \tag{4.17}$$

Adding and subtracting (4.16) and (4.17) we have

$$a \cdot b = \frac{1}{2}(ab + ba) \tag{4.18}$$

$$a \wedge b = \frac{1}{2}(ab - ba) \tag{4.19}$$

which are useful relationships, and are employed frequently.

We can also represent (4.19) as a determinant:

$$a \wedge b = \frac{1}{2}\begin{vmatrix} a & a \\ b & b \end{vmatrix} \tag{4.20}$$

which is slightly more compact and helps us understand other outer products.

Now although it is obvious that (4.19) results directly from subtracting (4.16) and (4.17), we could have arrived at the same result using the following argument:

Given two vectors

$$a = a_1 e_1 + a_2 e_2 + a_3 e_3$$
$$b = b_1 e_1 + b_2 e_2 + b_3 e_3 \tag{4.21}$$

their dyad creates two sets of unit dyads: one with the form

$$e_i e_i \qquad [i = 1, 2, 3] \tag{4.22}$$

and the other with the form

$$e_i e_j \qquad [i, j = 1, 2, 3] \quad (i \neq j). \tag{4.23}$$

Accepting that the geometric product ab creates a particular pattern of unit dyads, the product ba causes no change in the $e_i e_i$ terms, but a sign reversal in every term of the $e_i e_j$ set. Thus, subtracting ba from ab removes the $e_i e_i$ set leaving behind the $e_i e_j$ set of terms repeated twice, that happen to be the terms of the outer product. Hence we can write

$$a \wedge b = \frac{1}{2}(ab - ba). \tag{4.24}$$

Therefore, could it be that $a \wedge b \wedge c$ has a similar structure such as

$$a \wedge b \wedge c = \frac{1}{\lambda} \begin{vmatrix} a & a & a \\ b & b & b \\ c & c & c \end{vmatrix} \quad [\lambda \in \mathbb{R}]? \tag{4.25}$$

Expanding the determinant (4.25) gives

$$a \wedge b \wedge c = \frac{1}{\lambda}[(abc - cba) + (bca - acb) + (cab - bac)] \tag{4.26}$$

where λ is some integer.

To discover whether this relationship holds, all that we need do is apply the above argument to this configuration of vectors. So, given

$$a = a_1 e_1 + a_2 e_2 + a_3 e_3$$
$$b = b_1 e_1 + b_2 e_2 + b_3 e_3$$
$$c = c_1 e_1 + c_2 e_2 + c_3 e_3 \tag{4.27}$$

it is obvious that the triple product abc contains 27 terms involving every combination of a_i, b_i and c_i, each, of which, is associated with a triple of e_i terms.

Each triple must be in one of three possible sets:
three terms of the form

$$\pm e_1 e_2 e_3 \tag{4.28}$$

three terms of the form

$$e_i e_i e_i \quad [i = 1, 2, 3] \tag{4.29}$$

or all the other permutations of e_1, e_2 and e_3 apart from the above sets.

From a geometric product viewpoint, any trivector terms in the first and third sets are irreducible. The $e_i e_i e_i$ terms reduce to e_i (a vector) and the remaining terms must also reduce to a vector:

$$\pm e_i \quad [i = 1, 2, 3]. \tag{4.30}$$

There can be no bivector terms.

Thus we have shown that the geometric product of three vectors gives rise to a single trivector term and possibly various vector terms.

Assuming that antisymmetric pairs are the key to the solution, as it was in the 2D case, there can only be 3 such combinations in the 3D scenario:

$$(abc \,\&\, cba), \quad (bca \,\&\, acb), \quad (cab \,\&\, bac) \tag{4.31}$$

and the only possible basis vector combinations that arise are shown in Table 4.5.

TABLE 4.5   The triple basis vector combinations

Geometric Product		Result	Reverse Geometric Product		Result
$e_i \, e_i \, e_i$	$=$	e_i	$e_i \, e_i \, e_i$	$=$	e_i
$e_i \, e_i \, e_j$	$=$	e_j	$e_j \, e_i \, e_i$	$=$	e_j
$e_i \, e_j \, e_i$	$=$	$-e_j$	$e_i \, e_j \, e_i$	$=$	$-e_j$
$e_i \, e_j \, e_k$	$=$	$e_i \, e_j \, e_k$	$e_k \, e_j \, e_i$	$=$	$-e_i \, e_j \, e_k$

It is clear from Table 4.5 that when the geometric product is reversed, a sign change only occurs for trivector terms. This means that if we form the sum

$$(abc - cba) + (bca - acb) + (cab - bac). \tag{4.32}$$

we are left with six trivector terms, which permits us to write

$$a \wedge b \wedge c = \frac{1}{6}[(abc - cba) + (bca - acb) + (cab - bac)] \tag{4.33}$$

or

$$a \wedge b \wedge c = \frac{1}{6} \begin{vmatrix} a & a & a \\ b & b & b \\ c & c & c \end{vmatrix}. \tag{4.34}$$

The outer (exterior) product of n vectors $a_1, a_2, \ldots, a_n$ is often defined as the totally antisymmetrised sum of all geometric products of the vectors:

$$a_1 \wedge a_2 \ldots \wedge a_n = \frac{1}{n!} \sum (-1)^\varepsilon a_{k_1} a_{k_2} \ldots a_{k_n}. \tag{4.35}$$

The subscripts $k_1, k_2 \ldots, k_n$ represent every permutation of $1, 2, \ldots, n$, and $(-1)^\varepsilon$ is $+1$ or -1 as the permutation $k_1, k_2 \ldots, k_n$ is even or odd respectively.

"If two elements in a permutation of distinct elements are in reverse order relative to their normal or natural order, they constitute an inversion. Furthermore, a permutation is said to be even if it contains an even number of inversions; it is odd if the number of inversions is odd." [Jan Gullberg, p. 191]

Using Gullberg's definition:

For abc (0 inversions – i.e. even).
For bca b and c precede a (2 inversions – i.e. even).
For cab c precedes a and b (2 inversions – i.e. even).
For cba c precedes b and a, and b precedes a (3 inversions i.e. odd).
For acb c precedes b (1 inversion – i.e. odd).
For bac b precedes a (1 inversion – i.e. odd).

Therefore,

$$abc, bca, cab \text{ are even permutations, while}$$

$$cba, acb, bac \text{ are odd permutations.}$$

Thus in 2D, the permutations are

$$ab \text{ (even) and } ba \text{ (odd)}$$

which makes

$$a \wedge b = \frac{1}{2!}(ab - ba). \tag{4.36}$$

And in 3D, the permutations are

$$abc \text{ (even)}, acb \text{ (odd)}, bac \text{ (odd)}, bca \text{ (even)}, cab \text{ (even)}, cba \text{ (odd)}$$

which makes

$$a \wedge b \wedge c = \frac{1}{3!}[(abc - cba) + (bca - acb) + (cab - bac)]. \tag{4.37}$$

Now we can see that the 1/2! in (4.36) arises because there are only two permutations of ab, whereas the 1/6 in (4.33) arises because there are six permutations of abc, which means that we can rewrite both formulae in the form:

$$a \wedge b = \frac{1}{2!}(ab - ba) \tag{4.38}$$

$$a \wedge b \wedge c = \frac{1}{3!}[(abc - cba) + (bca - acb) + (cab - bac)]. \tag{4.39}$$

Appendix B shows all the terms for the six products in (4.39) and confirms the above reasoning.

4.4 Blades

The outer product of k vectors is called a *k-blade* with the following order:

$$vector = 1\text{-}blade$$

$$bivector = 2\text{-}blade$$

$$trivector = 3\text{-}blade, etc.$$

where k identifies the grade of a blade. Naturally, higher dimensions introduce new blades, but in general, the trivector is the highest blade required for every-day 3D Euclidean problems. Some authors include scalars within the above grouping and call them a 0-blade.

Describing vectors in terms of basis vectors brings order and elegance to vector analysis. Similarly, k-blades exhibit the same degree of order and elegance when they are described in terms of so called basis k-blades. Table 4.6 summarizes the 2D basis k-blades, and Table 4.7 the 3D basis k-blades.

TABLE 4.6   2D basis k-blades

k	basis k-blade	total
0 (scalar)	$\{1\}$	1
1 (vector)	$\{e_1, e_2\}$	2
2 (bivector)	$\{e_{12}\}$	1

TABLE 4.7   3D basis k-blades

k	basis k-blade	total
0 (scalar)	$\{1\}$	1
1 (vector)	$\{e_1, e_2, e_3\}$	3
2 (bivector)	$\{e_{12}, e_{23}, e_{31}\}$	3
3 (trivector)	$\{e_{123}\}$	1

There is no need to stop at three dimensions. In fact, by exploring 4D space confirms a pattern that is starting to emerge, so Table 4.8 shows the 4D basis k-blades.

The numbers in the 'total' column of each table are our old friend the binomial coefficients, and arise due to the possible ways for combining increasing numbers of vectors in twos, threes and fours, etc. The formula for generating the binomial coefficients is

$$\binom{n}{k} = \frac{n!}{(n-k)!k!} \tag{4.40}$$

TABLE 4.8   4D basis k-blades

k	basis k-blade	total
0 (scalar)	$\{1\}$	1
1 (vector)	$\{e_1, e_2, e_3, e_4\}$	4
2 (bivector)	$\{e_{12}, e_{13}, e_{14}, e_{23}, e_{24}, e_{34}\}$	6
3 (trivector)	$\{e_{123}, e_{124}, e_{134}, e_{234}\}$	4
4 (quadvector)	$\{e_{1234}\}$	1

where n is the total number of things to chose from, and k is the number of things chosen at any one time. By definition, $0! = 1$, therefore

$$\binom{n}{0} = \frac{n!}{(n - 0)!0!} = \frac{n!}{n!} = 1. \tag{4.41}$$

For example, in the 2D case, we have 2 vectors, which provide two ways of selecting a single vector, and one way of selecting a bivector:

$$\binom{2}{1} = \frac{2!}{(2 - 1)!1!} = 2$$

$$\binom{2}{2} = \frac{2!}{(2 - 2)!2!} = 1. \tag{4.42}$$

In the 3D case, we have 3 vectors, which provide three ways of selecting a single vector, and three ways of selecting a bivector, and one way of selecting a trivector:

$$\binom{3}{1} = \frac{3!}{(3 - 1)!1!} = 3$$

$$\binom{3}{2} = \frac{3!}{(3 - 2)!2!} = 3$$

$$\binom{3}{3} = \frac{3!}{(3 - 3)!3!} = 1. \tag{4.43}$$

The binomial coefficient formula (4.40) is used to compute the precise number of basis k-blades in a space of any arbitrary dimension.

Let us now explore how the geometric product behaves with various vector combinations.

4.5 The geometric product of different vectors

We begin with two vectors a and b whose sum is c:

$$c = a + b. \tag{4.44}$$

Therefore,

$$c^2 = (a+b)^2 = a^2 + b^2 + ab + ba. \tag{4.45}$$

Equation (4.45) contains some interesting features that are revealed when different vector combinations are introduced. For example, a and b can be orthogonal, linearly dependent, or linearly independent. Let's start with orthogonal vectors.

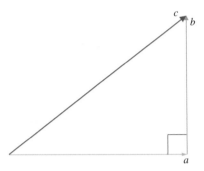

FIGURE 4.1.

4.5.1 Orthogonal vectors

Figure 4.1 shows vectors a and b where the relationship

$$c^2 = a^2 + b^2 + ab + ba \tag{4.46}$$

must hold. Furthermore, we know that

$$|c|^2 = |a|^2 + |b|^2. \tag{4.47}$$

But the modulus axiom allows us to write (4.47) as

$$c^2 = a^2 + b^2 \tag{4.48}$$

which forces us to conclude that

$$ab + ba = 0 \tag{4.49}$$

or put another way

$$ab = -ba \tag{4.50}$$

which confirms that orthogonal vectors anticommute.

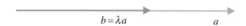

$$b = \lambda a \qquad\qquad a$$

FIGURE 4.2.

4.5.2 Parallel vectors

Figure 4.2 shows vectors a and b where

$$b = \lambda a \qquad\qquad [\lambda \in \mathbb{R}]. \qquad\qquad (4.51)$$

Applying the above axioms, the product ab is

$$ab = a\lambda a = \lambda aa = ba \qquad\qquad (4.52)$$

furthermore,

$$\lambda aa = \lambda a^2 = \lambda |a|^2 \qquad\qquad (4.53)$$

which confirms that parallel vectors commute, and that the product ab is a scalar.

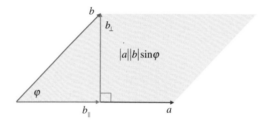

FIGURE 4.3.

4.5.3 Linearly independent vectors

Figure 4.3 shows two vectors a and b where vector b is a combination of components parallel and perpendicular to a:

$$b = b_{\parallel} + b_{\perp}. \qquad\qquad (4.54)$$

Therefore, we can write

$$ab = a(b_{\parallel} + b_{\perp})$$
$$= ab_{\parallel} + ab_{\perp}. \qquad\qquad (4.55)$$

As $b_{\parallel}$ is parallel with a, the product $ab_{\parallel}$ is

$$ab_{\parallel} = a \cdot b_{\parallel} + a \wedge b_{\parallel} = a \cdot b_{\parallel} \qquad\qquad (4.56)$$

therefore, $ab_{\parallel}$ is the inner product component.

Next, as $b_\perp$ is orthogonal to a, the product $ab_\perp$ is

$$ab_\perp = a \cdot b_\perp + a \wedge b_\perp = a \wedge b_\perp \qquad (4.57)$$

therefore, $ab_\perp$ is the outer product component.

Consequently, the geometric product of two linearly independent vectors is the sum of the inner and outer products formed by

$$ab = a \cdot b_\parallel + a \wedge b_\perp. \qquad (4.58)$$

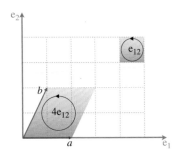

FIGURE 4.4.

Let's illustrate this relationship with the scenario shown in Figure 4.4. The vectors are

$$a = 2e_1$$
$$b = e_1 + 2e_2 \qquad (4.59)$$

therefore, their geometric product is

$$
\begin{aligned}
ab &= a \cdot b + a \wedge b \\
&= 2e_1 \cdot (e_1 + 2e_2) + 2e_1 \wedge (e_1 + 2e_2) \\
&= (2e_1 \cdot e_1 + 4e_1 \cdot e_2) + (2e_1 \wedge e_1 + 4e_1 \wedge e_2) \\
&= 2 + 4e_1 \wedge e_2 \\
&= 2 + 4e_{12}. \qquad (4.60)
\end{aligned}
$$

The 2 in (4.60) represents the scalar product, whilst the 4 represents the signed area of the parallelogram located on the plane containing e_1 and e_2. The fact that the 4 is positive implies that the vectors a and b are oriented in the same sense as the basis bivector e_{12}. For example, reversing the vectors to

$$a = e_1 + 2e_2$$
$$b = 2e_1 \qquad (4.61)$$

creates this geometric product:

$$ab = a \cdot b + a \wedge b$$
$$= (e_1 + 2e_2) \cdot 2e_1 + (e_1 + 2e_2) \wedge 2e_1$$
$$= 2e_1 \cdot e_1 + 4e_2 \cdot e_1 + 2e_1 \wedge e_1 + 4e_2 \wedge e_1$$
$$= 2 + 4e_2 \wedge e_1$$
$$= 2 - 4e_1 \wedge e_2$$
$$= 2 - 4e_{12}. \tag{4.62}$$

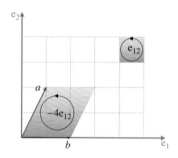

FIGURE 4.5.

The scalar term remains unchanged, whereas the area is negative, which indicates that the orientation of the two vectors opposes that of the basis bivector e_{12}. Figure 4.5 shows this change in orientation.

The area-calculating feature of the outer product invites us to solve geometric problems on an area basis, and we explore some of these problems in later chapters. For the moment, though, consider the scenario illustrated in Figure 4.6.

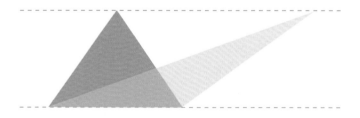

FIGURE 4.6.

Simple geometry reminds us that if the two horizontal lines are parallel, then two triangles sharing the same base and vertical height have equal areas. Similarly, in Figure 4.7, the two parallelograms must also have equal areas.

Therefore, the parallelograms in Figure 4.8 must also have identical areas and is readily confirmed by the outer product.

FIGURE 4.7.

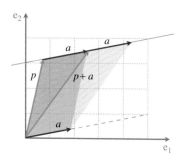

FIGURE 4.8.

If we form the outer product $p \wedge a$ we create the parallelogram shaded green. But note what happens if we place vector a further along the line and form the outer product of $p + a$ and a:

$$(p + a) \wedge a. \tag{4.63}$$

As the algebra is distributive over addition, we have

$$(p + a) \wedge a = p \wedge a + a \wedge a = p \wedge a. \tag{4.64}$$

Thus as we slide vector a along the line, the area of the parallelogram is constant. This geometric feature is quite useful in solving all sorts of problems.

4.6 Summary

The geometric product of two vectors is the sum of the inner and outer products, and is represented by the complete dyadic expansion of the two vectors. Hopefully, the tables provide a useful way to visualise the terms that arise in the various expansions.

In the next chapter we see how a new geometric algebra emerges, where scalars, vectors, bivectors, trivectors and higher dimensional objects are manipulated simultaneously. This is an algebra that has been hidden for many years because Grassmann's ideas were not adopted.

5 Geometric Algebra

5.1 Introduction

In this chapter we investigate various features of geometric algebra's products: the inner, outer and geometric products. One such feature is the ability to combine scalars, vectors, bivectors, trivectors, etc., to create a *multivector*, which can be treated just like any ordinary algebraic expression. We also discover that bivectors and trivectors are imaginary, in that they square to -1, which means that we are always working with a complex algebra. However, this in no way makes geometric algebra more difficult that any other algebra. So let's begin this investigation by introducing grades and pseudoscalars.

5.2 Grades and pseudoscalars

In the algebraic systems explored so far, we have considered the products of scalars, complex numbers, quaternions and vectors. In geometric algebra we have discovered that the outer product of vectors creates two new elements: a bivector and a trivector, which happen to be imaginary. And this is just the beginning! In order to bring some degree of order to these elements, the term *grade* is introduced.

Scalars are defined as grade 0, vectors grade 1, bivectors grade 2, and trivectors grade 3, and so on for higher dimensions. In such graded algebras it is traditional to call the highest grade element a *pseudoscalar*. Therefore, e_{12} is the 2D pseudoscalar and e_{123} is the 3D pseudoscalar. Later on we will see that pseudoscalars possess some interesting properties.

Our algebraic expressions now start to include scalars, vectors, bivectors and trivectors, and in order to isolate a particular grade from an expression we use the following notation:

$$\langle \text{expression} \rangle_g \tag{5.1}$$

where g identifies a particular grade. For example, say we have the following expression:

$$2 + 3e_1 + 2e_2 - 5e_{12} + 6e_{123} \tag{5.2}$$

J. Vince, *Geometric Algebra: An Algebraic System for Computer Games and Animation*,
© Springer-Verlag London Limited 2009

we extract the scalar term using:

$$\langle 2 + 3e_1 + 2e_2 - 5e_{12} + 6e_{123} \rangle_0 = 2 \tag{5.3}$$

the vector term using:

$$\langle 2 + 3e_1 + 2e_2 - 5e_{12} + 6e_{123} \rangle_1 = 3e_1 + 2e_2 \tag{5.4}$$

the bivector term using:

$$\langle 2 + 3e_1 + 2e_2 - 5e_{12} + 6e_{123} \rangle_2 = -5e_{12} \tag{5.5}$$

and the trivector term using:

$$\langle 2 + 3e_1 + 2e_2 - 5e_{12} + 6e_{123} \rangle_3 = 6e_{123}. \tag{5.6}$$

It is also worth pointing out that the inner vector product converts two grade 1 elements, i.e. vectors, into a grade 0 element, i.e. a scalar, whereas the outer vector product converts two grade 1 elements into a grade 2 element, i.e. a bivector. Thus the inner product is a grade lowering operation, while the outer product is a grade raising operation. These qualities of the inner and outer products continue as we deal with higher grade elements in the algebra. This is why the scalar product was renamed as the inner product, because the scalar product is synonymous with transforming vectors into scalars. Whereas, the inner product transforms two elements of grade n into a grade $n - 1$ element.

5.3 Multivectors

A *multivector* is a new mathematical structure comprising a linear combination of elements of different grades, such as scalars, vectors, bivectors, trivectors, etc. For example, in 2D the following is a multivector

$$1 + 2e_1 + 3e_2 + 4e_{12} \tag{5.7}$$

so, too, are

$$3$$

$$3 + e_1$$

$$2e_1$$

$$4e_1 + 5e_2$$

$$2 - 8e_{12}$$

$$4 + 6e_1 - 2e_{12}. \tag{5.8}$$

We are implying that an element of any grade can be linearly combined with any other element, which permits us to evolve a powerful algebra. For example, given

$$A = 2 + 3e_1 + 4e_2 + 8e_{12}$$

$$B = 1 + 2e_1 + 2e_2 + 6e_{12} \tag{5.9}$$

then

$$A + B = 3 + 5e_1 + 6e_2 + 14e_{12} \tag{5.10}$$

and

$$A - B = 1 + e_1 + 2e_2 + 2e_{12} \tag{5.11}$$

and using the geometric product

$$
\begin{aligned}
AB &= (2 + 3e_1 + 4e_2 + 8e_{12})(1 + 2e_1 + 2e_2 + 6e_{12}) \\
&= 2 + 4e_1 + 4e_2 + 12e_{12} \\
&\quad + 3e_1 + 6e_1^2 + 6e_{12} + 18e_{112} \\
&\quad + 4e_2 + 8e_{21} + 8e_2^2 + 24e_{212} \\
&\quad + 8e_{12} + 16e_{121} + 16e_{122} + 48e_{1212} \\
&= 2 + 4e_1 + 4e_2 + 12e_{12} \\
&\quad + 3e_1 + 6 + 6e_{12} + 18e_2 \\
&\quad + 4e_2 - 8e_{12} + 8 - 24e_1 \\
&\quad + 8e_{12} - 16e_2 + 16e_1 - 48
\end{aligned}
$$

and

$$AB = -32 - 3e_1 + 10e_2 + 18e_{12}. \tag{5.12}$$

Although we rarely use such expressions, they can be created.

We can prove that the algebra of 2D multivectors is closed by expanding a general product. If the product only contains terms of the original multivectors, then the algebra is closed. This is shown in Table 5.1.

TABLE 5.1   The geometric product AB

$A = a_0 + a_1e_1 + a_2e_2 + a_{12}e_{12}$ $B = b_0 + b_1e_1 + b_2e_2 + b_{12}e_{12}$			
AB	b_0	b_1e_1	b_2e_2	$b_{12}e_{12}$
a_0	a_0b_0	$a_0b_1e_1$	$a_0b_2e_2$	$a_0b_{12}e_{12}$
a_1e_1	$a_1b_0e_1$	a_1b_1	$a_1b_2e_{12}$	$a_1b_{12}e_2$
a_2e_2	$a_2b_0e_2$	$-a_2b_1e_{12}$	a_2b_2	$-a_2b_{12}e_1$
$a_{12}e_{12}$	$a_{12}b_0e_{12}$	$-a_{12}b_1e_2$	$a_{12}b_2e_1$	$-a_{12}b_{12}$

We can see from Table 5.1 that the product entries only contain terms found in the original multivectors, therefore, the algebra is closed. Furthermore, if we let

$$AB = C = \lambda_0 + \lambda_1e_1 + \lambda_2e_2 + \lambda_{12}e_{12} \tag{5.13}$$

then

$$\lambda_0 = a_0 b_0 + a_1 b_1 + a_2 b_2 - a_{12} b_{12}$$

$$\lambda_1 = a_0 b_1 + a_1 b_0 + a_{12} b_2 - a_2 b_{12}$$

$$\lambda_2 = a_0 b_2 + a_2 b_0 + a_1 b_{12} - a_{12} b_1$$

$$\lambda_{12} = a_0 b_{12} + a_{12} b_0 + a_1 b_2 - a_2 b_1. \tag{5.14}$$

A 3D multivector contains one extra vector term, two extra bivector terms, and a trivector term. Table 5.2 shows the product of two such multivectors, where it is clear from the table entries that the algebra is also closed.

TABLE 5.2   The geometric product AB

$A = a_0 + a_1 e_1 + a_2 e_2 + a_3 e_3 + a_{12} e_{12} + a_{23} e_{23} + a_{31} e_{31} + a_{123} e_{123}$ $B = b_0 + b_1 e_1 + b_2 e_2 + b_3 e_3 + b_{12} e_{12} + b_{23} e_{23} + b_{31} e_{31} + b_{123} e_{123}$							
AB	b_0	$b_1 e_1$	$b_2 e_2$	$b_3 e_3$	$b_{12} e_{12}$	$b_{23} e_{23}$	$b_{31} e_{31}$	$b_{123} e_{123}$
a_0	$a_0 b_0$	$a_0 b_1 e_1$	$a_0 b_2 e_2$	$a_0 b_3 e_3$	$a_0 b_{12} e_{12}$	$a_0 b_{23} e_{23}$	$a_0 b_{31} e_{31}$	$a_0 b_{123} e_{123}$
$a_1 e_1$	$a_1 b_0 e_1$	$a_1 b_1$	$a_1 b_2 e_{12}$	$-a_1 b_3 e_{31}$	$a_1 b_{12} e_2$	$a_1 b_{23} e_{123}$	$-a_1 b_{31} e_3$	$a_1 b_{123} e_{23}$
$a_2 e_2$	$a_2 b_0 e_2$	$-a_2 b_1 e_{12}$	$a_2 b_2$	$a_2 b_3 e_{23}$	$-a_2 b_{12} e_1$	$a_2 b_{23} e_3$	$a_2 b_{31} e_{123}$	$a_2 b_{123} e_{31}$
$a_3 e_3$	$a_3 b_0 e_3$	$a_3 b_1 e_{31}$	$-a_3 b_2 e_{23}$	$a_3 b_3$	$a_3 b_{12} e_{123}$	$-a_3 b_{23} e_2$	$a_3 b_{31} e_1$	$a_3 b_{123} e_{12}$
$a_{12} e_{12}$	$a_{12} b_0 e_{12}$	$-a_{12} b_1 e_2$	$a_{12} b_2 e_1$	$a_{12} b_3 e_{123}$	$-a_{12} b_{12}$	$-a_{12} b_{23} e_{31}$	$a_{12} b_{31} e_{23}$	$-a_{12} b_{123} e_3$
$a_{23} e_{23}$	$a_{23} b_0 e_{23}$	$a_{23} b_1 e_{123}$	$-a_{23} b_2 e_3$	$a_{23} b_3 e_2$	$a_{23} b_{12} e_{31}$	$-a_{23} b_{23}$	$-a_{23} b_{31} e_{12}$	$-a_{23} b_{123} e_1$
$a_{31} e_{31}$	$a_{31} b_0 e_{31}$	$a_{31} b_1 e_3$	$a_{31} b_2 e_{123}$	$-a_{31} b_3 e_1$	$-a_{31} b_{12} e_{23}$	$a_{31} b_{23} e_{12}$	$-a_{31} b_{31}$	$-a_{31} b_{123} e_2$
$a_{123} e_{123}$	$a_{123} b_0 e_{123}$	$a_{123} b_1 e_{23}$	$a_{123} b_2 e_{31}$	$a_{123} b_3 e_{12}$	$-a_{123} b_{12} e_3$	$-a_{123} b_{23} e_1$	$-a_{123} b_{31} e_2$	$-a_{123} b_{123}$

Another important feature of the geometric product is that it is associative:

$$A(BC) = (AB)C = ABC \tag{5.15}$$

and distributive over addition:

$$A(B + C) = AB + AC \tag{5.16}$$

where $A, B, \ldots, C$ are multivectors containing elements of arbitrary grade.

5.4 Reversion

You will have noticed how sensitive geometric algebra is to the sequence of vectors. Therefore it should not be too much of a surprise to learn that a special command is used to reverse the sequence of elements within an expression. For example, given

$$A = ab \tag{5.17}$$

the reverse is denoted using the dagger symbol $A^\dagger$

$$A^\dagger = ba \tag{5.18}$$

or the tilde $\tilde{A}$

$$\tilde{A} = ba. \tag{5.19}$$

Obviously, scalars are unaffected by reversion, neither are vectors. However, basis bivectors and trivectors flip their signs:

$$(e_1e_2)^\dagger = e_2e_1 = -e_1e_2 \tag{5.20}$$

and

$$(e_1e_2e_3)^\dagger = e_3e_2e_1 = -e_1e_2e_3. \tag{5.21}$$

The reader should be prepared to encounter both notations within the technical literature:

$$(e_1e_2)^\sim = e_2e_1 = -e_1e_2 \tag{5.22}$$

and

$$(e_1e_2e_3)^\sim = e_3e_2e_1 = -e_1e_2e_3. \tag{5.23}$$

Although bivectors and trivectors flip signs when reversed, the next pair of blades do not:

$$\tilde{e}_{1234} = e_{1234}$$

$$\tilde{e}_{12345} = e_{12345} \tag{5.24}$$

and this pattern of pairs continues for higher dimensions. A formula that switches signs in this manner is:

$$\tilde{e}_k = (-1)^{\frac{k(k-1)}{2}} e_k \tag{5.25}$$

where k is the grade of the blade.

When reversing a multivector containing terms up to a trivector, it's only the bivector and trivector terms that are reversed:

$$A = \lambda + a + B + T \qquad [\lambda \in \mathbb{R}] \tag{5.26}$$

where

λ is a scalar

a is a vector

B is a bivector, and

T is a trivector.

Then

$$\tilde{A} = \lambda + a - B - T \tag{5.27}$$

Let's illustrate this reversion process with an example.

Given three vectors

$$a = 2e_1 + 3e_2$$

$$b = 4e_1 - 2e_2$$

$$c = e_1 + e_2 \tag{5.28}$$

the product

$$ab = (2e_1 + 3e_2)(4e_1 - 2e_2) = 2 - 16e_{12} \tag{5.29}$$

and

$$ba = (4e_1 - 2e_2)(2e_1 + 3e_2) = 2 + 16e_{12}. \tag{5.30}$$

Thus

$$(ab)^\dagger = ba \tag{5.31}$$

which is confirmed by (5.26) and (5.27) where we observe that

$$(2 - 16e_{12})^\dagger = 2 + 16e_{12}. \tag{5.32}$$

Furthermore, if we expand the product abc we obtain

$$abc = (2e_1 + 3e_2)(4e_1 - 2e_2)(e_1 + e_2)$$

$$= (2 - 16e_{12})(e_1 + e_2)$$

$$= -14e_1 + 18e_2. \tag{5.33}$$

And as there are only vector terms, the reverse product cba will not change any signs:

$$cba = (e_1 + e_2)(4e_1 - 2e_2)(2e_1 + 3e_2)$$

$$= (2 - 6e_{12})(2e_1 + 3e_2)$$

$$= -14e_1 + 18e_2. \tag{5.34}$$

5.5 The inverse of a multivector

In linear algebra we often classify $n \times n$ matrices into two types: singular and non-singular. For example, a non-singular matrix A is invertible if there exists an $n \times n$ matrix A^{-1} such that $AA^{-1} = A^{-1}A = I_n$, where I_n is the identity matrix. Otherwise, A is called singular and non-invertible.

Well, it just so happens that geometric algebra also has a matrix representation and therefore opens the door to singular (non-invertible) matrices, which in turn means that as multivectors have a matrix form, not all can be inverted. Nevertheless, given a multivector A it will generally have a left inverse such that

$$A_L^{-1}A = I_n \tag{5.35}$$

and a right inverse such that

$$AA_R^{-1} = I_n. \tag{5.36}$$

One type of multivector that lends itself for inversion has the form

$$A = a_1 a_2 a_3 \ldots a_n \tag{5.37}$$

where $a_1 a_2 a_3 \ldots a_n$ are vectors, and A is their collective geometric product. Such multivectors are called *versors*.

The reverse of the versor (5.37) is

$$A^\dagger = a_n \ldots a_3 a_2 a_1. \tag{5.38}$$

Now let's see what happens when we multiply $A^\dagger$ by A:

$$A^\dagger A = (a_n \ldots a_3 a_2 a_1)(a_1 a_2 a_3 \ldots a_n). \tag{5.39}$$

Removing the parentheses, the associative quality of the geometric product ensures that pairs of vectors transform into scalars:

$$\begin{aligned} A^\dagger A &= (a_n \ldots (a_3(a_2(a_1 a_1)a_2)a_3) \ldots a_n). \\ &= |a_1|^2 (a_n \ldots (a_3(a_2 a_2)a_3) \ldots a_n) \\ &= |a_1|^2 + |a_2|^2 + |a_3|^2 \ldots + |a_n|^2. \end{aligned} \tag{5.40}$$

Furthermore,

$$\begin{aligned} AA^\dagger &= (a_1 a_2 a_3 \ldots a_n)(a_n \ldots a_3 a_2 a_1) \\ &= |a_1|^2 + |a_2|^2 + |a_3|^2 \ldots + |a_n|^2 \end{aligned}$$

and

$$A^\dagger A = AA^\dagger. \tag{5.41}$$

As we require that

$$A^{-1}A = 1 \tag{5.42}$$

we can write

$$\begin{aligned} (A^{-1}A)A^\dagger &= A^\dagger \\ A^{-1}(AA^\dagger) &= A^\dagger \end{aligned} \tag{5.43}$$

and

$$A^{-1} = \frac{A^\dagger}{AA^\dagger}. \tag{5.44}$$

But we already know that $AA^\dagger$ is a scalar, which means that the inverse of a versor is its reversion divided by $AA^\dagger$ or $A^\dagger A$.

Therefore, we have

$$A^{-1}A = \frac{A^\dagger}{A^\dagger A}A = \frac{A^\dagger A}{A^\dagger A} = 1 \tag{5.45}$$

or

$$AA^{-1} = A\frac{A^\dagger}{AA^\dagger} = \frac{AA^\dagger}{AA^\dagger} = 1 \qquad (5.46)$$

which implies that there is no difference between A_L^{-1} and A_R^{-1}.

Using (5.44), if we let

$$A = b$$
$$A^\dagger = b \qquad (5.47)$$

and

$$AA^\dagger = |b|^2 \qquad (5.48)$$

then

$$A^{-1} = b^{-1} = \frac{b}{|b|^2} \qquad (5.49)$$

Therefore, given ab, we can recover vector a by post-multiplying by b^{-1}:

$$a = abb^{-1}$$
$$= ab\frac{b}{|b|^2}. \qquad (5.50)$$

Similarly, we can recover vector b by pre-multiplying by a^{-1}:

$$b = a^{-1}ab$$
$$= \frac{a}{|a|^2}ab. \qquad (5.51)$$

For example, given two vectors

$$a = e_1 + 2e_2$$
$$b = 3e_1 + 2e_2 \qquad (5.52)$$

their geometric product is

$$ab = (e_1 + 2e_2) \cdot (3e_1 + 2e_2) + (e_1 + 2e_2) \wedge (3e_1 + 2e_2)$$
$$= 7 - 4e_1 \wedge e_2$$
$$= 7 - 4e_{12}. \qquad (5.53)$$

Using (5.51)

$$b = \frac{e_1 + 2e_2}{5}(7 - 4e_{12})$$
$$= \frac{1}{5}(7e_1 - 4e_{112} + 14e_2 - 8e_{212})$$
$$= \frac{1}{5}(7e_1 - 4e_2 + 14e_2 + 8e_1)$$
$$= 3e_1 + 2e_2 \qquad (5.54)$$

which is the vector b.

Similarly, given ab and b, a is retrieved as follows:

$$a = (7 - 4e_{12})\frac{3e_1 + 2e_2}{13}$$

$$= \frac{1}{13}(21e_1 + 14e_2 - 12e_{121} - 8e_{122})$$

$$= \frac{1}{13}(21e_1 + 14e_2 + 12e_2 - 8e_1)$$

$$= e_1 + 2e_2. \tag{5.55}$$

Note that because we cannot assume that the geometric product is commutative we must ensure that the inverse vector is on the correct side of the product. Now let's demonstrate that inversion works for 3D vectors. For example, give the following vectors:

$$a = e_1 + 2e_2 + 3e_3$$

$$b = 2e_1 - e_2 + 2e_3 \tag{5.56}$$

then

$$ab = (e_1 + 2e_2 + 3e_3)(2e_1 - e_2 + 2e_3)$$

$$= 6 + (e_1 + 2e_2 + 3e_3) \wedge (2e_1 - e_2 + 2e_3)$$

$$= 6 - e_{12} - 2e_{31} - 4e_{12} + 4e_{23} + 6e_{31} + 3e_{23}$$

$$= 6 - 5e_{12} + 7e_{23} + 4e_{31}. \tag{5.57}$$

But

$$b = \frac{a}{|a|^2}ab$$

$$= \frac{1}{14}(e_1 + 2e_2 + 3e_3)(6 - 5e_{12} + 7e_{23} + 4e_{31})$$

$$= \frac{1}{14}\left(\begin{matrix} 6e_1 - 5e_2 + 7e_{123} - 4e_3 + 12e_2 + 10e_1 + 14e_3 \\ + 8e_{123} + 18e_3 - 15e_{123} - 21e_2 + 12e_1 \end{matrix} \right)$$

$$= \frac{1}{14}(28e_1 - 14e_2 + 28e_3)$$

$$= 2e_1 - e_2 + 2e_3$$

$$= b. \tag{5.58}$$

It is left to the reader to invert a multivector.

Note that the inverse of a unit vector is the original vector:

$$\hat{a}^{-1} = \frac{\hat{a}}{|\hat{a}|^2} = \hat{a}. \tag{5.59}$$

5.6 The imaginary properties of the outer product

So far we know that the outer product of two vectors is represented by one or more unit basis bivectors, such as

$$a \wedge b = \lambda_1 e_{12} + \lambda_2 e_{23} + \lambda_3 e_{31} \tag{5.60}$$

where, in this case, the λ_i terms represent areas projected onto their respective unit basis bivectors. But what has not emerged is that the outer product is an imaginary quantity, which is revealed by expanding e_{12}^2:

$$e_{12}^2 = e_{1212} \tag{5.61}$$

but as

$$e_{21} = -e_{12} \tag{5.62}$$

then

$$e_{1(21)2} = -e_{1(12)2}$$
$$= -e_1^2 e_2^2$$
$$= -1 \tag{5.63}$$

thus

$$(e_1 \wedge e_2)^2 = -1. \tag{5.64}$$

Consequently, the geometric product effectively creates a complex number! Thus in a 2D scenario, given two vectors

$$a = a_1 e_1 + a_2 e_2$$
$$b = b_1 e_1 + b_2 e_2 \tag{5.65}$$

their geometric product is

$$ab = (a_1 b_1 + a_2 b_2) + (a_1 b_2 - a_2 b_1) e_{12} \tag{5.66}$$

and knowing that $e_{12} = i$, then we can write ab as

$$ab = (a_1 b_1 + a_2 b_2) + (a_1 b_2 - a_2 b_1) i. \tag{5.67}$$

However, this notation is not generally adopted by the geometric algebra community. The reason being that i is normally only associated with a scalar, with which it commutes. Whereas in 2D, e_{12} is associated with scalars and vectors, and although scalars present no problem, under some conditions, it anticommutes with vectors. Consequently, an upper-case I is used so that there is no confusion between the two elements. Thus (5.67) is written as

$$ab = (a_1 b_1 + a_2 b_2) + (a_1 b_2 - a_2 b_1) I \tag{5.68}$$

where

$$I^2 = -1. \tag{5.69}$$

5.7 The rotational properties of the 2D unit bivector

Geometric algebra offers a rather elegant way to perform rotations, which is very reminiscent of quaternions, which, in turn, are reminiscent of complex numbers. Just to recap, given a complex number

$$z = a + ib \tag{5.70}$$

when this is repeatedly multiplied by i we obtain the sequence:

$$i(a + ib) = -b + ia$$

$$i(-b + ia) = -a - ib$$

$$i(-a - ib) = b - ia$$

$$i(b - ia) = a + ib. \tag{5.71}$$

After four multiplications by i we return to the original complex number, which when drawn on the complex plane shows that at each stage the complex number is rotated anticlockwise 90°, as shown in Figure 5.1. Therefore, it should not be any surprise that the pseudoscalar performs the same operation.

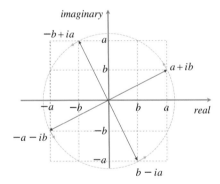

FIGURE 5.1.

As an example, given the vector

$$a_1e_1 + a_2e_2 \tag{5.72}$$

when this is repeatedly multiplied by I (i.e. e_{12}) we obtain the sequence:

$$e_{12}(a_1e_1 + a_2e_2) = a_1e_{121} + a_2e_{122}$$

$$= a_2e_1 - a_1e_2$$

$$e_{12}(a_2e_1 - a_1e_2) = a_2e_{121} - a_1e_{122}$$

$$= -a_1e_1 - a_2e_2$$

$$e_{12}(-a_1e_1 - a_2e_2) = -a_1e_{121} - a_2e_{122}$$

$$= -a_2e_1 + a_1e_2$$

$$e_{12}(-a_2e_1 + a_1e_2) = -a_2e_{121} + a_1e_{122}$$

$$= a_1e_1 + a_2e_2. \tag{5.73}$$

Once again, after four multiplications we return to the original vector. But has the vector been rotated clockwise or anticlockwise? Well, Figure 5.2 confirms that it has been rotated clockwise!

If we want to rotate the vector anticlockwise, we must to post-multiply the vector by the pseudoscalar:

$$(a_1e_1 + a_2e_2)e_{12} = a_1e_{112} + a_2e_{212}$$

$$= -a_2e_1 + a_1e_2$$

$$(-a_2e_1 + a_1e_2)e_{12} = -a_2e_{112} + a_1e_{212}$$

$$= -a_1e_1 - a_2e_2$$

$$(-a_1e_1 - a_2e_2)e_{12} = -a_1e_{112} - a_2e_{212}$$

$$= a_2e_1 - a_1e_2$$

$$(a_2e_1 - a_1e_2)e_{12} = a_2e_{112} - a_1e_{212}$$

$$= a_1e_1 + a_2e_2. \tag{5.74}$$

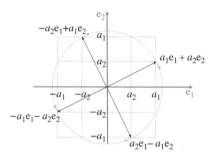

FIGURE 5.2.

Therefore, the pseudoscalar anticommutes with vectors in 2D:

$$aI = -Ia. \tag{5.75}$$

5.8 The imaginary properties of the 3D unit bivector and the trivector

Having discovered that the 2D unit bivector is an imaginary quantity, in that it squares to -1, it should be no surprise that the 3D unit bivectors also share the same imaginary property, and is readily shown by squaring the following terms:

$$e_{12}^2 = e_{1212} = -e_{1221} = -1. \tag{5.76}$$

Similarly,

$$e_{23}^2 = e_{2323} = -e_{2332} = -1 \tag{5.77}$$

and

$$e_{31}^2 = e_{3131} = -e_{3113} = -1. \tag{5.78}$$

Similarly, expanding

$$e_{123}^2 = e_{123123}$$

$$= e_{1212}$$

$$= -e_{1221}$$

$$= -1. \tag{5.79}$$

Therefore, the unit trivector is also an imaginary, and to distinguish between 2D and 3D pseudoscalars I will employ the following notation:

$$I_{12} \text{ for 2D} \tag{5.80}$$

and

$$I_{123} \text{ for 3D.} \tag{5.81}$$

5.9 Duality

In 1825, Joseph Gergonne [1771–1859], identified the principle of duality where he observed that when the term *point* is substituted for *line*, *lie on* for *pass through*, *collinear* for *concurrent*, *intersection* for *join*, or vice versa, results in an equally valid theorem which is the "dual" of the original. For example, "two points lie on a line" has a dual "two intersecting lines pass through a point". More loosely, one can argue that "two lines define a plane" or its dual "two planes intersect at a line", or "three points form a plane", or the dual "three planes intersect at a point".

The ability to exchange pairs of geometric elements such as lines and planes involves a dual operation, which in geometric algebra is relatively easy to define. For example, given a multivector A, its dual A^* is defined as

$$A^* = IA \tag{5.82}$$

where I is the pseudoscalar representing the highest blade in the algebra used for A. For two and three-dimensional space this will be either e_{12} or e_{123}, which means that

$$I_{12} = e_{12} \tag{5.83}$$

and

$$I_{123} = e_{123}. \tag{5.84}$$

Therefore, given the 2D vector

$$a = a_1e_1 + a_2e_2 \tag{5.85}$$

its dual is

$$a^* = e_{12}(a_1 e_1 + a_2 e_2)$$

$$= a_1 e_{121} + a_2 e_{122}$$

$$= a_2 e_1 - a_1 e_2. \tag{5.86}$$

Which is another vector rotated 90° anticlockwise.

If we take the dual of a^* we obtain a further 90° rotation anticlockwise:

$$(a^*)^* = e_{12}(a_2 e_1 - a_1 e_2)$$

$$= a_2 e_{121} - a_1 e_{122}$$

$$= -(a_1 e_1 + a_2 e_2) = -a. \tag{5.87}$$

And taking the dual of $(a^*)^*$ we obtain a further 90° rotation anticlockwise:

$$\left((a^*)^*\right)^* = -e_{12}(a_1 e_1 + a_2 e_2)$$

$$= -(a_1 e_{121} + a_2 e_{122})$$

$$= -(a_2 e_1 - a_1 e_2) = -a^*. \tag{5.88}$$

Finally, taking the dual of $((a^*)^*)^*$ we return to the original vector a:

$$\left(\left((a^*)^*\right)^*\right)^* = -e_{12}(a_2 e_1 - a_1 e_2)$$

$$= -(a_2 e_{121} - a_1 e_{122})$$

$$= a_1 e_1 + a_2 e_2 = a. \tag{5.89}$$

These duals are shown in Figure 5.3.

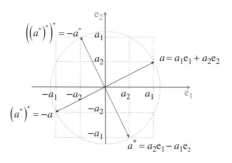

FIGURE 5.3.

In 3D, the dual of vector e_1 is

$$e_{123} e_1 = e_{1231} = e_{23}. \tag{5.90}$$

Similarly, the dual of e_2 is

$$e_{123}e_2 = e_{1232} = e_{31} \tag{5.91}$$

and the dual of e_3 is

$$e_{123}e_3 = e_{1233} = e_{12}. \tag{5.92}$$

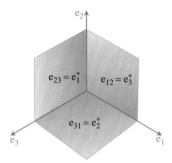

FIGURE 5.4.

These duals are shown in Figure 5.4.

Note that the duals of these vector are perpendicular bivectors. For a general vector $a_1e_1 + a_2e_2 + a_3e_3$ its dual is

$$e_{123}(a_1e_1 + a_2e_2 + a_3e_3) = a_1e_{1231} + a_2e_{1232} + a_3e_{1233}$$
$$= a_3e_{12} + a_1e_{23} + a_2e_{31}. \tag{5.93}$$

Now let's find the duals of the 3D basis bivectors:

$$e_{123}e_{12} = e_{12312} = -e_3$$
$$e_{123}e_{23} = e_{12323} = -e_1$$
$$e_{123}e_{31} = e_{12331} = -e_2. \tag{5.94}$$

Therefore, the dual of a unit basis bivector is the negative perpendicular axial vector, and are shown in Figure 5.5.

Having discovered the duals of vectors and bivectors, now let's observe the cyclic pattern caused by repeating the dual operation to successive results. We begin by taking the dual of e_1:

$$Ie_1 = e_{123}e_1 = e_{23}. \tag{5.95}$$

Next we take the dual of e_{23}:

$$Ie_{23} = e_{123}e_{23} = -e_1 \tag{5.96}$$

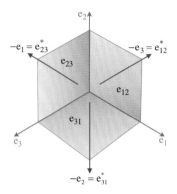

FIGURE 5.5.

followed by the dual of $-e_1$:

$$I(-e_1) = -e_{123}e_1 = -e_{23} \tag{5.97}$$

followed by the dual of $-e_{23}$:

$$I(-e_{23}) = -e_{123}e_{23} = e_1 \tag{5.98}$$

which brings us back to the starting point. Similar patterns arise with the other basis vectors.

5.10 Summary

Some very important ideas have been introduced in this chapter which will be employed in later chapters. In particular, grades, pseudoscalars, multivectors, duals, inversion, and reversion will be used to solve all sorts of geometric problems. The imaginary nature of bivectors and trivectors does not introduce any unwanted problems. What is interesting though, is that just like imaginary i, a bivector can also rotate vectors through 90°.

In the next two chapters we explore the 2D and 3D products that arise between scalars, vectors, bivectors, and trivectors.

6 Products in 2D

6.1 Introduction

In this chapter we examine the products that arise when combining scalars, vectors and the bivector. In some cases the results are trivial, or have been touched upon in previous chapters, but some of the products are new and require investigation. Before we start, let's remind ourselves of the axioms behind geometric algebra:

Axiom 1: The associative rule

$$a(bc) = (ab)c \qquad (6.1)$$

Axiom 2: The left and right distributive rules

$$a(b + c) = ab + ac$$

$$(b + c)a = ba + ca \qquad (6.2)$$

Axiom 3: $\qquad (\lambda a)b = \lambda(ab) = \lambda ab \qquad [\lambda \in \mathbb{R}] \qquad (6.3)$

Axiom 4: $\qquad \lambda(\varepsilon a) = (\lambda \varepsilon)a \qquad [\lambda, \varepsilon \in \mathbb{R}] \qquad (6.4)$

Axiom 5: $\qquad \lambda(a + b) = \lambda a + \lambda b \qquad [\lambda \in \mathbb{R}] \qquad (6.5)$

Axiom 6: $\qquad (\lambda + \varepsilon)a = \lambda a + \varepsilon a \qquad [\lambda, \varepsilon \in \mathbb{R}] \qquad (6.6)$

Axiom 7: $\qquad a^2 = |a|^2 \qquad (6.7)$

J. Vince, *Geometric Algebra: An Algebraic System for Computer Games and Animation*, © Springer-Verlag London Limited 2009

Axiom 8: $$|a \cdot b| = |a||b| \cos \theta \qquad (6.8)$$

Axiom 9: $$|a \wedge b| = |a||b| \sin \theta \qquad (6.9)$$

Axiom 10: $$ab = a \cdot b + a \wedge b \qquad (6.10)$$

Axiom 11: $$a \wedge b = -b \wedge a. \qquad (6.11)$$

The objective of this chapter is to tabulate the inner, outer and geometric products of vectors and the 2D bivector. But before we begin let's quickly review how scalars interact with vectors and bivectors.

6.2 The scalar-vector product

Multiplying a vector by a scalar can modify the vector's magnitude and orientation. The vector's magnitude is resized by the scalar's magnitude, whilst its orientation is reversed if the scalar's sign is negative. This is demonstrated as follows:

Given a vector

$$a = a_1 e_1 + a_2 e_2 \qquad (6.12)$$

and a scalar

$$\lambda \qquad [\lambda \in \mathbb{R}] \qquad (6.13)$$

then

$$\lambda a = \lambda a_1 e_1 + \lambda a_2 e_2. \qquad (6.14)$$

The orientation of a is given by θ where

$$\theta = \cos^{-1}(a_1/|a|), \qquad (6.15)$$

and the orientation of λa is given by θ' where

$$\theta' = \cos^{-1}(\pm \lambda a_1/|\lambda a|) = \pm \theta. \qquad (6.16)$$

Therefore, when a vector is multiplied by a scalar, a change of orientation occurs if the scalar is negative.

The magnitude of a is given by

$$|a| = \sqrt{a_1^2 + a_2^2}, \qquad (6.17)$$

and the magnitude of λa is given by

$$|\lambda a| = \sqrt{\lambda^2 a_1^2 + \lambda^2 a_2^2}$$

$$= |\lambda| \sqrt{a_1^2 + a_2^2}$$

$$= |\lambda||a|. \qquad (6.18)$$

Therefore, when a vector is multiplied by a scalar, its length is scaled by the scalar's magnitude.

6.3 The scalar-bivector product

Given a bivector

$$B = \beta e_{12} \tag{6.19}$$

and a scalar

$$\lambda \qquad [\lambda \in \mathbb{R}] \tag{6.20}$$

then

$$\lambda B = \lambda \beta e_{12} \tag{6.21}$$

which effectively scales the area associated with the original bivector. Note that if λ is negative, it reverses the bivector's orientation.

6.4 The vector-vector products

The product of two vectors is fundamental to geometric algebra as it establishes a new interpretation of the inner product, a new outer product, and the geometric product, which is the sum of the inner and outer products. At the same time, the cross product is quietly dropped, as if it had never existed!

6.4.1 The inner product

Geometric algebra embraces the 2D scalar product, but renames it as the inner product. This aligns it with the inner products involving vectors, bivectors and trivectors, which are grade-lowering products. The scalar product has always been a grade-lowering product, in that it converts two vectors into a scalar, but this quality has never been its major distinction. The scalar product is generally used to test two vectors for orthogonality or to reveal their separating angle.

Let's begin by computing the inner product of two arbitrary vectors using geometric algebra:
Given two vectors

$$a = a_1 e_1 + a_2 e_2$$
$$b = b_1 e_1 + b_2 e_2 \tag{6.22}$$

then

$$a \cdot b = (a_1 e_1 + a_2 e_2) \cdot (b_1 e_1 + b_2 e_2)$$
$$= a_1 b_1 e_1 \cdot e_1 + a_1 b_2 e_1 \cdot e_2 + a_2 b_1 e_2 \cdot e_1 + a_2 b_2 e_2 \cdot e_2$$
$$= a_1 b_1 + a_2 b_2. \tag{6.23}$$

From a geometric algebra perspective the inner product (the old scalar product) is defined as

$$a \cdot b = \frac{1}{2}(ab + ba). \tag{6.24}$$

We have already discovered in chapter 4 that the inner product of two vectors is symmetric while the outer product is antisymmetric. Thus when we add ab to ba the outer-product terms cancel, and the inner-product terms are repeated twice – hence the $\frac{1}{2}$ in (6.24).

To determine $a \cdot b$ we need to compute ab and ba:

$$ab = (a_1 b_1 + a_2 b_2) + (a_1 b_2 - a_2 b_1)e_{12}$$
$$ba = (a_1 b_1 + a_2 b_2) - (a_1 b_2 - a_2 b_1)e_{12}. \qquad (6.25)$$

Then using (6.24) we have

$$a \cdot b = a_1 b_1 + a_2 b_2. \qquad (6.26)$$

The axiom that gives rise to the grade-lowering quality of the inner product is

$$a \cdot b = |a||b| \cos \theta \qquad (6.27)$$

which ensures that

$$e_1^2 = e_2^2 = 1 \qquad (6.28)$$

and

$$e_1 \cdot e_2 = e_2 \cdot e_1 = 0. \qquad (6.29)$$

Another meaning for (6.26) is

$$|a||b| \cos \theta = a_1 b_1 + a_2 b_2 \qquad (6.30)$$

where θ is the angle between the vectors. This means that

$$a \cdot b = |a||b| \cos \theta = a_1 b_1 + a_2 b_2 \qquad (6.31)$$

and provides a mechanism to determine θ for two vectors using:

$$\theta = \cos^{-1} \left(\frac{a_1 b_1 + a_2 b_2}{|a||b|} \right). \qquad (6.32)$$

Table 6.1 shows the inner products of the basis vectors. And although this seems a trivial result to record, it forms part of a larger table that describes the entire algebra. Table 6.2 records the inner products of two arbitrary vectors a and b.

TABLE 6.1   Inner product of the 2D basis vectors

Inner product	e_1	e_2
e_1	1	
e_2		1

TABLE 6.2   Inner product of two 2D vectors

Inner product	a	b
a	$|a|^2$	$a \cdot b$
b	$a \cdot b$	$|b|^2$

6.4.2 The outer product

The outer product of two vectors is defined as

$$a \wedge b = \frac{1}{2}(ab - ba). \tag{6.33}$$

We also know that

$$|a \wedge b| = |a||b| \sin \theta \tag{6.34}$$

where θ is the angle between the two vectors.

As the outer product is antisymmetric and the inner product is symmetric, when we subtract ba from ab the inner-product terms cancel, and the outer-product terms are repeated twice – hence the $\frac{1}{2}$ in (6.33).

To determine $a \wedge b$ we need to compute ab and ba (6.33):

therefore,

$$a \wedge b = (a_1 b_2 - a_2 b_1)e_{12}. \tag{6.35}$$

The scalar term $(a_1 b_2 - a_2 b_1)$ represents the area of a parallelogram associated with the plane containing the two basis vectors e_1 and e_2. The shape of the area is immaterial – what is important is its sign. When it is positive, it implies that the vectors are in the same directional sense as the reference bivector e_{12}, i.e. clockwise. When it is negative, it implies that the sense of the vectors opposes that of the bivector. When it is zero, it implies that the vectors are parallel, and the outer product cannot form an area.

The antisymmetric nature of the outer product reminds us that

$$a \wedge b = -b \wedge a. \tag{6.36}$$

Furthermore, as

$$e_1 \cdot e_2 = 0 \tag{6.37}$$

then

$$e_1 e_2 = e_1 \wedge e_2 = e_{12}. \tag{6.38}$$

This permits us to substitute e_{12} for $e_1 \wedge e_2$, or any similar combination.

Table 6.3 shows the outer products for the two basis vectors, and Table 6.4 shows the outer products for two arbitrary vectors a and b.

TABLE 6.3   Outer product of the 2D basis vectors

Outer product	e_1	e_2
e_1		e_{12}
e_2	$-e_{12}$	

TABLE 6.4   Outer product of two 2D vectors

Outer product	a	b
a		$a \wedge b$
b	$-a \wedge b$	

6.4.3 The geometric product

The geometric product of two vectors is defined as

$$ab = a \cdot b + a \wedge b \tag{6.39}$$

which is the algebraic expansion of the dyadic formed by ab. The geometric product is obviously very sensitive to the vectors used. For example:

with identical vectors:

$$aa = a \cdot a + a \wedge a = |a|^2 \tag{6.40}$$

with parallel vectors:

$$a(\lambda a) = a \cdot (\lambda a) + a \wedge (\lambda a) = \lambda |a|^2 \tag{6.41}$$

and with orthogonal vectors where $|a| = |a_\perp|$:

$$a(\lambda a_\perp) = a \cdot (\lambda a_\perp) + a \wedge (\lambda a_\perp) = \lambda |a|^2 \, e_{12}. \tag{6.42}$$

Table 6.5 shows the geometric products for the two basis vectors, and Table 6.6 shows the geometric products of two arbitrary vectors a and b.

TABLE 6.5   Geometric product of the 2D basis vectors

Geometric product	e_1	e_2
e_1	1	e_{12}
e_2	$-e_{12}$	1

Next, we investigate what happens when we multiply vectors and bivectors together.

TABLE 6.6   Geometric product of two 2D vectors

Geometric product	a	b
a	$\lvert a \rvert^2$	$a \cdot b + a \wedge b$
b	$a \cdot b - a \wedge b$	$\lvert b \rvert^2$

6.5 The vector-bivector product

In the previous chapter we discovered that the unit basis bivector is imaginary, in that it squares to -1, and like imaginary i possesses rotational properties. But unlike imaginary i, the unit bivector anticommutes with its vector partner. For example, given

$$a = a_1 e_1 + a_2 e_2 \qquad (6.43)$$

then

$$e_{12} a = a_2 e_1 - a_1 e_2 \qquad (6.44)$$

but

$$a e_{12} = -a_2 e_1 + a_1 e_2. \qquad (6.45)$$

Which shows that they anticommute. Furthermore, pre-multiplying a vector by the unit bivector rotates the vector 90° clockwise, while post-multiplying rotates the vector 90° anticlockwise. Note that the operation is a grade-lowering operation, which implies that it results from the action of the inner product, and that the outer product is zero. The outer product has to be zero as there is no space orthogonal to the bivector.

Table 6.7 summarizes these results, which at the end of the chapter are integrated with the other products.

TABLE 6.7   Geometric product of the 2D basis vectors and bivector

Geometric product	e_1	e_2	e_{12}
e_1	1	e_{12}	e_2
e_2	$-e_{12}$	1	$-e_1$
e_{12}	$-e_2$	e_1	-1

Therefore, for a general bivector B

$$Ba = -aB \qquad (6.46)$$

and

$$aB = a \cdot B + a \wedge B = a \cdot B \qquad (6.47)$$

and

$$Ba = B \cdot a + B \wedge a = -a \cdot B \tag{6.48}$$

which permits us to write

$$a \cdot B = \frac{1}{2}(aB - Ba). \tag{6.49}$$

For example, given

$$a = a_1 e_1 + a_2 e_2 \tag{6.50}$$

and

$$B = \lambda e_{12} \tag{6.51}$$

then

$$aB = (a_1 e_1 + a_2 e_2)\lambda e_{12} = \lambda(-a_2 e_1 + a_1 e_2) \tag{6.52}$$
$$Ba = \lambda e_{12}(a_1 e_1 + a_2 e_2) = \lambda(a_2 e_1 - a_1 e_2) \tag{6.53}$$

and

$$a \cdot B = \frac{\lambda}{2}[(-a_2 e_1 + a_1 e_2) - (a_2 e_1 - a_1 e_2)]$$
$$a \cdot B = \lambda(-a_2 e_1 + a_1 e_2) \tag{6.54}$$

which confirms that the vector has been rotated anticlockwise 90° and scaled by λ and corresponds with (6.45).

6.6 The bivector-bivector product

The unit bivector is a pseudoscalar and squares to -1, which gives it its imaginary qualitities:

$$e_{12}e_{12} = e_{1212} = -e_{1221} = -1. \tag{6.55}$$

When using arbitrary bivectors αe_{12} and βe_{12}, their product is

$$\alpha e_{12}\beta e_{12} = -\alpha\beta. \tag{6.56}$$

However, is it the inner or the outer product that gives rise to this result? Well, as it is a grade-lowering operation, it must be the inner product. Furthermore, the outer product has no extra space to perform its work, and must return a null result. Therefore:

$$e_{12} \cdot e_{12} = -1$$
$$e_{12} \wedge e_{12} = 0 \tag{6.57}$$

and

$$e_{12}^2 = e_{12} \cdot e_{12} + e_{12} \wedge e_{12} = -1. \tag{6.58}$$

One additional product converts a vector a into a complex number z by pre-multiplying it by e_1:
given

$$a = a_1 e_1 + a_2 e_2 \tag{6.59}$$

then

$$z = e_1 a$$
$$= e_1(a_1 e_1 + a_2 e_2)$$
$$= a_1 + a_2 e_{12}$$
$$z = a_1 + a_2 i. \tag{6.60}$$

Reversing the product produces:

$$a e_1 = (a_1 e_1 + a_2 e_2) e_1$$
$$= a_1 - a_2 e_{12}$$
$$= a_1 - a_2 i$$
$$a e_1 = z^* \tag{6.61}$$

where z^* is the complex conjugate of z.

We know from complex analysis that

$$z z^* = a_1^2 + a_2^2 \tag{6.62}$$

which is confirmed by computing

$$z z^* = e_1 a a e_1$$
$$= a^2$$
$$z z^* = a_1^2 + a_2^2. \tag{6.63}$$

6.7 Summary

In 2D we can only form products between vectors and bivectors, which restricts the range of results. Table 6.8 combines all the products from the above tables. What is rather interesting is that the outer product only accounts for two of the entries – the rest are the work of the inner product.

TABLE 6.8   Geometric product of the 2D basis vector and bivector

Geometric product	e_1	e_2	e_{12}
e_1	1	e_{12}	e_2
e_2	$-e_{12}$	1	$-e_1$
e_{12}	$-e_2$	e_1	-1

Tables 6.9, 6.10 and 6.11 show the inner, outer and geometric products for arbitrary scalars, vectors and bivectors.

TABLE 6.9   Inner product of a 2D vector and bivector

Inner product	b [vector]	B [bivector]
a [vector]	$a \cdot b$ [scalar]	$a \cdot B$ [vector] a is rotated 90° in B
A [bivector]	$A \cdot b$ [vector] b is rotated −90° in A	$-\|A\|\|B\|$ [scalar]

TABLE 6.10   Outer product of a 2D vector and bivector

Outer product	b [vector]	B [bivector]
a [vector]	$a \wedge b$ [bivector]	
A [bivector]		

TABLE 6.11   Geometric product of a 2D vector and bivector

Geometric product	b [vector]	B [bivector]
a [vector]	$a \cdot b + a \wedge b$ [multivector]	$a \cdot B$ [vector] a is rotated 90° in B
A [bivector]	$A \cdot b$ [vector] b is rotated −90° in A	$-\|A\|\|B\|$ [scalar]

The above summary should provide the reader with a good overview of how geometric algebra behaves in two dimensions. We now repeat the same exercise in three dimensions, which although is more detailed, is not more complex.

7 Products in 3D

7.1 Introduction

In this chapter we examine the products that arise when combining scalars, vectors, bivectors and trivectors in 3D. Some of the products have been covered in earlier chapters, but there are some new ideas that require explanation. We begin with the simple scalar-vector product.

7.2 The scalar-vector product

Multiplying a vector by a scalar can modify the vector's magnitude and orientation. The vector's magnitude is resized by the scalar's magnitude, whilst its orientation is reversed if the scalar's sign is negative. This is demonstrated as follows:

Given a vector

$$a = a_1 e_1 + a_2 e_2 + a_3 e_3 \tag{7.1}$$

and a scalar

$$\lambda \qquad [\lambda \in \mathbb{R}] \tag{7.2}$$

then

$$\lambda a = \lambda a_1 e_1 + \lambda a_2 e_2 + \lambda a_3 e_3. \tag{7.3}$$

The orientation of a is given by three direction cosine angles:

$$\alpha = \cos^{-1}(a_1/|a|)$$
$$\beta = \cos^{-1}(a_2/|a|)$$
$$\chi = \cos^{-1}(a_3/|a|) \tag{7.4}$$

J. Vince, *Geometric Algebra: An Algebraic System for Computer Games and Animation*,
© Springer-Verlag London Limited 2009

and the orientation of λa is given by three similar direction cosine angles:

$$\alpha' = \cos^{-1}(\pm\lambda a_1/|\lambda a|) = \pm\alpha$$
$$\beta' = \cos^{-1}(\pm\lambda a_2/|\lambda a|) = \pm\beta$$
$$\chi' = \cos^{-1}(\pm\lambda a_3/|\lambda a|) = \pm\chi. \tag{7.5}$$

Therefore, when a vector is multiplied by a scalar, a change of orientation occurs if the scalar is negative.

The magnitude of a is given by

$$|a| = \sqrt{a_1^2 + a_2^2 + a_3^2} \tag{7.6}$$

and the magnitude of λa is given by

$$|\lambda a| = \sqrt{\lambda^2 a_1^2 + \lambda^2 a_2^2 + \lambda^2 a_3^2}$$
$$= |\lambda|\sqrt{a_1^2 + a_2^2 + a_3^2}$$
$$= |\lambda||a|. \tag{7.7}$$

Therefore, when a vector is multiplied by a scalar, its length is scaled by the scalar's magnitude.

7.3 The scalar-bivector product

Multiplying a bivector by a scalar can modify the bivector's magnitude and orientation. The bivector's magnitude is resized by the scalar's magnitude, whilst its orientation is reversed if the scalar's sign is negative. This is demonstrated as follows:

Given a bivector

$$B = \alpha e_{12} + \beta e_{23} + \chi e_{31} \qquad [(\alpha, \beta, \chi) \in \mathbb{R}] \tag{7.8}$$

and a scalar

$$\lambda \qquad\qquad [\lambda \in \mathbb{R}] \tag{7.9}$$

then

$$\lambda B = \lambda\alpha e_{12} + \lambda\beta e_{23} + \lambda\chi e_{31} \tag{7.10}$$

which effectively scales the areas associated with the three bivectors. If λ is negative, it reverses the orientation of all the bivector areas.

7.4 The scalar-trivector product

Multiplying a trivector by a scalar can modify the trivector's magnitude and orientation. The trivector's magnitude is resized by the scalar's magnitude, whilst its orientation is reversed if the scalar's sign is negative. This is demonstrated as follows:

Given a trivector

$$\Psi e_{123} \qquad [\Psi \in \mathbb{R}] \tag{7.11}$$

and a scalar

$$\lambda \qquad [\lambda \in \mathbb{R}] \tag{7.12}$$

then

$$\lambda \Psi e_{123} \tag{7.13}$$

which scales the volume associated with the trivector. If λ is negative, it reverses the orientation of the trivector volume.

7.5 The vector-vector products

For completeness, we repeat some of the ideas covered in 6.4 to reinforce the fact that the outer product functions in space of any dimension, but can only create new geometric elements if the local space permits.

7.5.1 The inner product

The inner product is generally used to test two vectors for orthogonality or to reveal their separating angle. Therefore, given two vectors

$$a = a_1 e_1 + a_2 e_2 + a_3 e_3$$

$$b = b_1 e_1 + b_2 e_2 + b_3 e_3 \tag{7.14}$$

then

$$a \cdot b = (a_1 e_1 + a_2 e_2 + a_3 e_3) \cdot (b_1 e_1 + b_2 e_2 + b_3 e_3)$$

$$= a_1 b_1 + a_2 b_2 + a_3 b_3. \tag{7.15}$$

From a geometric algebra perspective the inner product is defined as

$$a \cdot b = \frac{1}{2}(ab + ba). \tag{7.16}$$

To determine $a \cdot b$ we need to compute ab and ba:

$$ab = (a_1 b_1 + a_2 b_2 + a_3 b_3) + (a_1 b_2 - a_2 b_1)e_{12} + (a_2 b_3 - a_3 b_2)e_{23} + (a_3 b_1 - a_1 b_3)e_{31}$$

$$ba = (a_1 b_1 + a_2 b_2 + a_3 b_3) - (a_1 b_2 - a_2 b_1)e_{12} - (a_2 b_3 - a_3 b_2)e_{23} - (a_3 b_1 - a_1 b_3)e_{31}. \tag{7.17}$$

Then using (7.16) we have

$$a \cdot b = a_1 b_1 + a_2 b_2 + a_3 b_3. \tag{7.18}$$

The axiom that gives rise to the grade-lowering quality of the inner product is

$$a \cdot b = |a||b| \cos \theta \tag{7.19}$$

which ensures that

$$e_1^2 = e_2^2 = e_3^2 = 1. \tag{7.20}$$

Another meaning for (7.18) is

$$|a||b| \cos \theta = a_1 b_1 + a_2 b_2 + a_3 b_3 \tag{7.21}$$

where θ is the angle between the vectors. This means that

$$a \cdot b = |a||b| \cos \theta = a_1 b_1 + a_2 b_2 + a_3 b_3 \tag{7.22}$$

and provides a mechanism to determine θ for two vectors using:

$$\theta = \cos^{-1} \left(\frac{a_1 b_1 + a_2 b_2 + a_3 b_3}{|a||b|} \right). \tag{7.23}$$

Table 7.1 shows the inner products of the basis vectors e_1, e_2, e_3, and Table 7.2 records the inner products of two vectors a and b.

TABLE 7.1   Inner product of the 3D basis vectors

Inner product	e_1	e_2	e_3
e_1	1		
e_2		1	
e_3			1

TABLE 7.2   Inner product of two 3D vectors

Inner product	a	b
a	$|a|^2$	$a \cdot b$
b	$a \cdot b$	$|b|^2$

7.5.2 The outer product

The outer product of two vectors is defined as

$$a \wedge b = \frac{1}{2}(ab - ba).\tag{7.24}$$

We also know that

$$|a \wedge b| = |a||b| \sin \theta \tag{7.25}$$

where θ is the angle between the two vectors.

To determine $a \wedge b$ we need ab and ba (7.17), therefore,

$$a \wedge b = (a_1 b_2 - a_2 b_1)e_{12} + (a_2 b_3 - a_3 b_2)e_{23} + (a_3 b_1 - a_1 b_3)e_{31}. \tag{7.26}$$

The scalar terms $(a_1 b_2 - a_2 b_1)$, $(a_2 b_3 - a_3 b_2)$, $(a_3 b_1 - a_1 b_3)$ represent the areas of the parallelograms projected onto the three planes defined by e_{12}, e_{23} and e_{31} respectively. Again, the shape of the areas is immaterial – what is important are their signs. A positive value implies that the projected vectors are in the same directional sense as the reference bivector, i.e. anticlockwise. A negative value implies that the sense of the projected vectors opposes that of the reference bivector. A zero value implies that the projected vectors are parallel, and the outer product cannot form an area.

The antisymmetric nature of the outer product reminds us that

$$a \wedge b = -b \wedge a. \tag{7.27}$$

Furthermore, as

$$e_1 \cdot e_2 = e_2 \cdot e_3 = e_3 \cdot e_1 = 0$$

we can employ the following substitutions:

$$e_1 \wedge e_2 = e_{12}$$
$$e_2 \wedge e_3 = e_{23}$$
$$e_3 \wedge e_1 = e_{31}. \tag{7.28}$$

Table 7.3 shows the outer products for the three basis vectors e_1, e_2, e_3, and Table 7.4 shows the outer products for two arbitrary vectors a and b.

TABLE 7.3   Outer product of the 3D basis vectors

Outer product	e_1	e_2	e_3
e_1		e_{12}	$-e_{31}$
e_2	$-e_{12}$		e_{23}
e_3	e_{31}	$-e_{23}$	

TABLE 7.4   Outer product of two 3D vectors

Outer product	a	b
a		$a \wedge b$
b	$-a \wedge b$	

7.5.3 The geometric product

The geometric product of two vectors is independent of the spatial dimension, therefore, the conditions described in section 6.4.3 still hold in 3D.

Table 7.5 shows the geometric products for the three basis vectors e_1, e_2, e_3 and Table 7.6 shows the geometric products of two arbitrary vectors a and b.

TABLE 7.5   Geometric product of the 3D basis vectors

Geometric product	e_1	e_2	e_3
e_1	1	e_{12}	$-e_{31}$
e_2	$-e_{12}$	1	e_{23}
e_3	e_{31}	$-e_{23}$	1

TABLE 7.6   Geometric product of two 3D vectors

Geometric product	a	b
a	$|a|^2$	$a \cdot b + a \wedge b$
b	$a \cdot b - a \wedge b$	$|b|^2$

7.6 The vector-bivector product

7.6.1 The dyadic approach

The dyadic approach to vector products shows that the geometric product is the sum of the inner and outer products, where given two vectors a and b:

$$ab = a \cdot b + a \wedge b \tag{7.29}$$

where

$$a \cdot b = \frac{1}{2}(ab + ba) \tag{7.30}$$

and

$$a \wedge b = \frac{1}{2}(ab - ba). \tag{7.31}$$

The same dyadic approach can be used to resolve the product cB where c is a vector and B is a bivector $a \wedge b$.

Therefore, given

$$a = a_1 e_1 + a_2 e_2 + a_3 e_3$$
$$b = b_1 e_1 + b_2 e_2 + b_3 e_3$$
$$c = c_1 e_1 + c_2 e_2 + c_3 e_3 \tag{7.32}$$

then

$$B = a \wedge b$$
$$= (a_1 e_1 + a_2 e_2 + a_3 e_3) \wedge (b_1 e_1 + b_2 e_2 + b_3 e_3)$$
$$= (a_1 b_2 - a_2 b_1)e_{12} + (a_2 b_3 - a_3 b_2)e_{23} + (a_3 b_1 - a_1 b_3)e_{31}.$$

$$\tag{7.33}$$

We now form the geometric products cB and Bc which are shown in Tables 7.7 and 7.8 respectively.

TABLE 7.7   Geometric product cB

	$c = c_1 e_1 + c_2 e_2 + c_3 e_3$ $B = a \wedge b = (a_1 e_1 + a_2 e_2 + a_3 e_3) \wedge (b_1 e_1 + b_2 e_2 + b_3 e_3)$		
cB	$(a_2 b_3 - a_3 b_2)e_{23}$	$(a_3 b_1 - a_1 b_3)e_{31}$	$(a_1 b_2 - a_2 b_1)e_{12}$
$c_1 e_1$	$c_1(a_2 b_3 - a_3 b_2)e_{123}$	$-c_1(a_3 b_1 - a_1 b_3)e_3$	$c_1(a_1 b_2 - a_2 b_1)e_2$
$c_2 e_2$	$c_2(a_2 b_3 - a_3 b_2)e_3$	$c_2(a_3 b_1 - a_1 b_3)e_{123}$	$-c_2(a_1 b_2 - a_2 b_1)e_1$
$c_3 e_3$	$-c_3(a_2 b_3 - a_3 b_2)e_2$	$c_3(a_3 b_1 - a_1 b_3)e_1$	$c_3(a_1 b_2 - a_2 b_1)e_{123}$

TABLE 7.8   Geometric product Bc

	$B = a \wedge b = (a_1 e_1 + a_2 e_2 + a_3 e_3) \wedge (b_1 e_1 + b_2 e_2 + b_3 e_3)$ $c = c_1 e_1 + c_2 e_2 + c_3 e_3$		
Bc	$c_1 e_1$	$c_2 e_2$	$c_3 e_3$
$(a_2 b_3 - a_3 b_2)e_{23}$	$c_1(a_2 b_3 - a_3 b_2)e_{123}$	$-c_2(a_2 b_3 - a_3 b_2)e_3$	$c_3(a_2 b_3 - a_3 b_2)e_2$
$(a_3 b_1 - a_1 b_3)e_{31}$	$c_1(a_3 b_1 - a_1 b_3)e_3$	$c_2(a_3 b_1 - a_1 b_3)e_{123}$	$-c_3(a_3 b_1 - a_1 b_3)e_1$
$(a_1 b_2 - a_2 b_1)e_{12}$	$-c_1(a_1 b_2 - a_2 b_1)e_2$	$c_2(a_1 b_2 - a_2 b_1)e_1$	$c_3(a_1 b_2 - a_2 b_1)e_{123}$

The inner-product terms are highlighted in yellow, while the outer-product terms are in blue. Tables 7.7 and 7.8 reveal that

$$c \wedge B = [c_1(a_2 b_3 - a_3 b_2) + c_2(a_3 b_1 - a_1 b_3) + c_3(a_1 b_2 - a_2 b_1)]e_{123} \tag{7.34}$$

$$c \cdot B = c_3(a_3 b_1 - a_1 b_3)e_1 + c_1(a_1 b_2 - a_2 b_1)e_2 + c_2(a_2 b_3 - a_3 b_2)e_3$$
$$- c_2(a_1 b_2 - a_2 b_1)e_1 - c_3(a_2 b_3 - a_3 b_2)e_2 - c_1(a_3 b_1 - a_1 b_3)e_3 \tag{7.35}$$

and

$$B \wedge c = [c_1(a_2 b_3 - a_3 b_2) + c_2(a_3 b_1 - a_1 b_3) + c_3(a_1 b_2 - a_2 b_1)]e_{123} \tag{7.36}$$

$$B \cdot c = -c_3(a_3 b_1 - a_1 b_3)e_1 - c_1(a_1 b_2 - a_2 b_1)e_2 - c_2(a_2 b_3 - a_3 b_2)e_3$$
$$+ c_2(a_1 b_2 - a_2 b_1)e_1 + c_3(a_2 b_3 - a_3 b_2)e_2 + c_1(a_3 b_1 - a_1 b_3)e_3. \tag{7.37}$$

Therefore, if we sum the terms cB and Bc the inner product terms cancel and we are left with twice the outer product terms, which permits us to write:

$$c \wedge B = B \wedge c = \frac{1}{2}(cB + Bc)$$

$$= [c_1(a_2b_3 - a_3b_2) + c_2(a_3b_1 - a_1b_3) + c_3(a_1b_2 - a_2b_1)]e_{123}$$

$$c \wedge B = B \wedge c = \begin{vmatrix} a_1 & b_1 & c_1 \\ a_2 & b_2 & c_2 \\ a_3 & b_3 & c_3 \end{vmatrix} e_{123}. \tag{7.38}$$

Similarly, subtracting Bc from cB, the outer product terms cancel and we are left with twice the inner product terms:

$$c \cdot B = -B \cdot c = \frac{1}{2}(cB - Bc)$$

$$= c_3(a_3b_1 - a_1b_3)e_1 + c_1(a_1b_2 - a_2b_1)e_2 + c_2(a_2b_3 - a_3b_2)e_3$$

$$- c_2(a_1b_2 - a_2b_1)e_1 - c_3(a_2b_3 - a_3b_2)e_2 - c_1(a_3b_1 - a_1b_3)e_3$$

$$= [c_3(a_3b_1 - a_1b_3) - c_2(a_1b_2 - a_2b_1)]e_1$$

$$+ [c_1(a_1b_2 - a_2b_1) - c_3(a_2b_3 - a_3b_2)]e_2$$

$$+ [c_2(a_2b_3 - a_3b_2) - c_1(a_3b_1 - a_1b_3)]e_3. \tag{7.39}$$

Equation (7.39) is concealing a hidden pattern which is revealed by writing it as the difference of two sets of terms as shown in Table 7.9:

TABLE 7.9   Product $c \cdot B$ (7.39) in tabular form

	$a_1c_1b_2e_2$	$a_1c_1b_3e_3$			$c_1b_1a_2e_2$	$c_1b_1a_3e_3$
$a_2c_2b_1e_1$		$a_2c_2b_3e_3$	$-$	$c_2b_2a_1e_1$		$c_2b_2a_3e_3$
$a_3c_3b_1e_1$	$a_3c_3b_2e_2$			$c_3b_3a_1e_1$	$c_3b_3a_2e_2$	

If we add the same diagonal terms to both sets, the equation is not disturbed – Table 7.10:

TABLE 7.10   Product $c \cdot B$ (7.39) with extra terms in tabular form

$a_1c_1b_1e_1$	$a_1c_1b_2e_2$	$a_1c_1b_3e_3$		$c_1b_1a_1e_1$	$c_1b_1a_2e_2$	$c_1b_1a_3e_3$
$a_2c_2b_1e_1$	$a_2c_2b_2e_2$	$a_2c_2b_3e_3$	$-$	$c_2b_2a_1e_1$	$c_2b_2a_2e_2$	$c_2b_2a_3e_3$
$a_3c_3b_1e_1$	$a_3c_3b_2e_2$	$a_3c_3b_3e_3$		$c_3b_3a_1e_1$	$c_3b_3a_2e_2$	$c_3b_3a_3e_3$

It now becomes clear from Table 7.10 that we have created:

$$c \cdot B = -B \cdot c = (c \cdot a)b - (c \cdot b)a. \tag{7.40}$$

If you feel that you have seen $(c \cdot a)b - (c \cdot b)a$ before, you are probably right, as it is the vector triple product:

$$(a \times b) \times c = (c \cdot a)b - (c \cdot b)a \tag{7.41}$$

which implies that

$$-B \cdot c \equiv (a \times b) \times c \equiv c \cdot B. \tag{7.42}$$

The resulting vector $(c \cdot a)b - (c \cdot b)a$ must be in the plane containing a and b i.e. the bivector B. What is not so obvious, is its relationship with c, which is something we discover in the next section.

7.6.2 The geometric approach

In section 5.7 we saw that the pseudoscalar e_{12} rotates vectors when it is used to multiply a vector. Let's now see if something similar happens in $\mathbb{R}^3$. We begin by defining a vector

$$a = a_1 e_1 + a_2 e_2 + a_3 e_3 \tag{7.43}$$

and a unit bivector

$$e_{12} \tag{7.44}$$

and constructing their product

$$e_{12}a = a_1 e_{12}e_1 + a_2 e_{12}e_2 + a_3 e_{12}e_3 \tag{7.45}$$

which becomes

$$= a_1 e_{121} + a_2 e_{122} + a_3 e_{123}$$

$$= -a_1 e_2 + a_2 e_1 + a_3 e_{123}$$

$$e_{12}a = a_2 e_1 - a_1 e_2 + a_3 e_{123}. \tag{7.46}$$

Equation (7.46) contains two parts:
a vector

$$a_2 e_1 - a_1 e_2 \tag{7.47}$$

and a volume

$$a_3 e_{123}. \tag{7.48}$$

Comparing (7.43) and (7.46) we see that the component of the vector contained within the bivector has been rotated clockwise 90° while the vector component orthogonal to the bivector has created a volume. Figure 7.1 illustrates how the projected vector is rotated within the plane e_{12}.

Post-multiplying the vector by the same bivector rotates the projected vector component in an anticlockwise direction:

$$ae_{12} = a_1 e_1 e_{12} + a_2 e_2 e_{12} + a_3 e_3 e_{12} \tag{7.49}$$

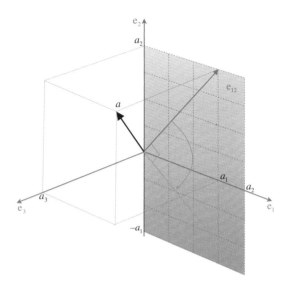

FIGURE 7.1.

which becomes

$$= a_1 e_{112} + a_2 e_{212} + a_3 e_{312}$$
$$= a_1 e_2 - a_2 e_1 + a_3 e_{123}$$
$$a e_{12} = -a_2 e_1 + a_1 e_2 + a_3 e_{123}. \tag{7.50}$$

Even though the vector has been rotated in an opposite direction, the volumetric component remains unaltered.

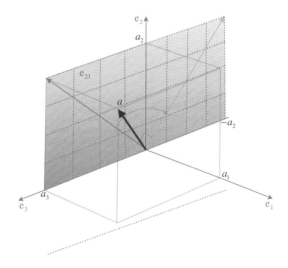

FIGURE 7.2.

We can almost predict what happens when the vector is multiplied by the other two bivectors. But just to make sure, let's perform the algebra, starting with e_{23}:

$$e_{23}a = a_1 e_{23}e_1 + a_2 e_{23}e_2 + a_3 e_{23}e_3$$

$$= a_1 e_{231} + a_2 e_{232} + a_3 e_{233}$$

$$= a_1 e_{123} - a_2 e_3 + a_3 e_2$$

$$e_{23}a = a_3 e_2 - a_2 e_3 + a_1 e_{123}. \tag{7.51}$$

This time, the vector's component contained in e_{23} is rotated 90°, while the vector component orthogonal to the bivector creates a volume. (Figure 7.2). Reversing the product reverses the direction of rotation. Similarly, for e_{31}:

$$e_{31}a = a_1 e_{31}e_1 + a_2 e_{31}e_2 + a_3 e_{31}e_3$$

$$= a_1 e_{311} + a_2 e_{312} + a_3 e_{313}$$

$$= a_1 e_3 + a_2 e_{123} - a_3 e_1$$

$$e_{31}a = a_1 e_3 - a_3 e_1 + a_2 e_1 e_{123}. \tag{7.52}$$

This time, the vector's component contained in e_{31} is rotated 90°, while the vector component orthogonal to the bivector creates a volume (Figure 7.3). Reversing the product reverses the direction of rotation.

But what happens when we form the product between a vector and a general bivector?

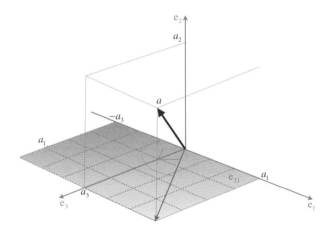

FIGURE 7.3.

To answer this question consider the scenario shown in Figure 7.4, where we see a bivector B and a vector a. We now investigate the geometric significance of the product aB.

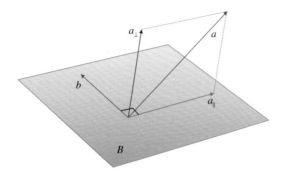

FIGURE 7.4.

The strategy is to resolve vector a into two components such that:

$$a = a_\parallel + a_\perp \qquad (7.53)$$

where

$$a_\parallel \text{ is parallel with the bivector } B$$

and

$$a_\perp \text{ is perpendicular to the bivector } B.$$

Therefore, we can write the product aB as

$$aB = (a_\parallel + a_\perp)B. \qquad (7.54)$$

The bivector B must be the outer product of two vectors that form its plane, which, by definition, must include $a_\parallel$. The second vector can be any other vector, so long as it lies in the same plane. And for convenience, but most of all because we know that it will be useful, let's assume that the second vector b is orthogonal to $a_\parallel$, as shown in Figure 7.4.

We can now write

$$B = a_\parallel \wedge b \qquad (7.55)$$

and because $a_\parallel$ is orthogonal to b

$$a_\parallel \cdot b = 0 \qquad (7.56)$$

therefore,

$$B = a_\parallel b. \qquad (7.57)$$

Next, we manipulate (7.41) algebraically to secure the desired result:

$$a_\parallel B = a_\parallel(a_\parallel b) = a_\parallel^2 b. \qquad (7.58)$$

The term $a_\parallel^2$ is a scalar, which means that $a_\parallel^2 b$ must be a vector.

Consequently, the product of vector $a_\parallel$ in the plane of bivector B is another vector $a_\parallel^2 b$, orthogonal to $a_\parallel$.

If we investigate the perpendicular component $a_\perp$, we discover that

$$a_\perp B = a_\perp(a_\parallel \wedge b) = a_\perp a_\parallel b \qquad (7.59)$$

which, from Figure 7.4, confirms $a_\perp, a_\parallel, b$ that are three orthogonal vectors that create a trivector.

Thus we have shown that, in general, the product of a vector and a bivector creates vector and trivector terms, where the resulting vector has been scaled and rotated anticlockwise 90° in the plane of the bivector.

As an example, Figure 7.5 shows a bivector B formed from two vectors

$$c = e_1$$
$$d = e_2 - e_3. \tag{7.60}$$

Therefore,

$$B = c \wedge d$$
$$= e_1 \wedge (e_2 - e_3)$$
$$= e_1 e_2 - e_1 e_3$$
$$= e_{12} + e_{31}. \tag{7.61}$$

If

$$a = e_1 \tag{7.62}$$

then the product aB is

$$aB = e_1(e_{12} + e_{31})$$
$$= e_2 - e_3. \tag{7.63}$$

Thus the vector a has been rotated anticlockwise 90° by the bivector B.

Note that the original vector a is a unit vector, but after rotation it is scaled by $\sqrt{2}$, which is the bivector's magnitude. Therefore, a unit bivector must be used to prevent arbitrary scaling.

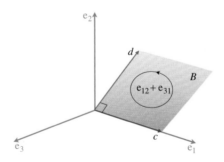

FIGURE 7.5.

Now let's examine the effect of multiplying vectors and bivectors by the 3D pseudoscalar.

7.7 The vector-trivector products

The trivector is the highest dimensional object that can exist in three dimensions, therefore, any attempt to form an outer product between a vector and a trivector must fail. Consequently, the products that do exist are the result of the inner product, and mean that the geometric product simply reflects these results.

7.7.1 The inner product

Tables 7.11 and 7.12 summarize the inner products between the basis vectors and the unit basis trivector. These products also represent the dual relationships between vectors and bivectors in 3D.

TABLE 7.11   Inner product of the 3D basis vectors and the pseudoscalar

Inner product	I_{123}
e_1	e_{23}
e_2	e_{31}
e_3	e_{12}

TABLE 7.12   Reverse inner product of the 3D basis vectors and the pseudoscalar

Inner product	e_1	e_2	e_3
I_{123}	e_{23}	e_{31}	e_{12}

The pseudoscalar commutes with all vectors in 3D, whereas in 2D it anticommutes. Observe how the product of a grade-3 pseudoscalar (I_{ijk}) with a grade-1 multivector (e_i) is a grade $(3-1)$ multivector (e_{jk}).

7.7.2 The outer product

The outer product of a vector and its pseudoscalar is zero.

7.7.3 The geometric product

The geometric product of a vector and its pseudoscalar is the same as the inner product, which are shown in Tables 7.13 and 7.14.

TABLE 7.13   Geometric product of the 3D basis vectors and the pseudoscalar

Geometric product	I_{123}
e_1	e_{23}
e_2	e_{31}
e_3	e_{12}

TABLE 7.14   Reverse geometric product of the 3D basis vectors and the pseudoscalar

Geometric product	e_1	e_2	e_3
I_{123}	e_{23}	e_{31}	e_{12}

Finally, Table 7.15 records the result of multiplying an arbitrary vector by the 3D pseudoscalar.

TABLE 7.15   Geometric product of the 3D pseudoscalar and vector

Geometric product	$a_1e_1 + a_2e_2 + a_3e_3$
I_{123}	$a_3e_{12} + a_1e_{23} + a_2e_{31}$

7.8 The bivector-bivector products

As only one basis bivector exists in 2D, the only product that can exist is a self product, which is imaginary:

$$e_{12}^2 = -1 \tag{7.64}$$

However, in 3D there are three orthogonal basis bivectors, some, of which, are pure imaginaries, null, and the others bivectors. The rules governing these products are as follows:

$$AB = A \cdot B + A \wedge B$$
$$BA = A \cdot B - A \wedge B \tag{7.65}$$

where A and B are bivectors.
    Therefore,

$$A \cdot B = \frac{1}{2}(AB + BA)$$

$$A \wedge B = \frac{1}{2}(AB - BA). \tag{7.66}$$

7.8.1 The inner product

Using the basis bivectors we have

$$e_{12} \cdot e_{12} = \frac{1}{2}(e_{12}e_{12} + e_{12}e_{12}) = -1 \tag{7.67}$$

consequently,

$$e_{23} \cdot e_{23} = e_{31} \cdot e_{31} = -1 \tag{7.68}$$

which meets with the grade-lowering qualities of the inner product.
   But we also have to consider the products:

$$e_{12} \cdot e_{23} = \tfrac{1}{2}(e_{12}e_{23} + e_{23}e_{12}) = 0$$
$$e_{12} \cdot e_{31} = \tfrac{1}{2}(e_{12}e_{31} + e_{31}e_{12}) = 0$$
$$e_{23} \cdot e_{31} = \tfrac{1}{2}(e_{23}e_{31} + e_{31}e_{23}) = 0$$
$$e_{23} \cdot e_{12} = \tfrac{1}{2}(e_{23}e_{12} + e_{12}e_{23}) = 0$$
$$e_{31} \cdot e_{12} = \tfrac{1}{2}(e_{31}e_{12} + e_{12}e_{31}) = 0$$
$$e_{31} \cdot e_{23} = \tfrac{1}{2}(e_{31}e_{23} + e_{23}e_{31}) = 0. \tag{7.69}$$

These inner products are summarized in Table 7.16.

TABLE 7.16   Inner product of the 3D basis bivector

Inner product	e_{23}	e_{31}	e_{12}
e_{23}	-1		
e_{31}		-1	
e_{12}			-1

7.8.2 The outer products

Using the basis bivectors we have

$$e_{12} \wedge e_{12} = \frac{1}{2}(e_{12}e_{12} - e_{12}e_{12}) = 0 \tag{7.70}$$

consequently,

$$e_{23} \wedge e_{23} = e_{31} \wedge e_{31} = 0. \tag{7.71}$$

Similarly, we also have to consider the products:

$$e_{12} \wedge e_{23} = \frac{1}{2}(e_{12}e_{23} - e_{23}e_{12}) = -e_{31}$$

$$e_{23} \wedge e_{12} = \frac{1}{2}(e_{23}e_{12} - e_{12}e_{23}) = e_{31}$$

$$e_{23} \wedge e_{31} = \frac{1}{2}(e_{23}e_{31} - e_{31}e_{23}) = -e_{12}$$

$$e_{31} \wedge e_{23} = \frac{1}{2}(e_{31}e_{23} - e_{23}e_{31}) = e_{12}$$

$$e_{31} \wedge e_{12} = \frac{1}{2}(e_{31}e_{12} - e_{31}e_{12}) = -e_{23}$$

$$e_{12} \wedge e_{31} = \frac{1}{2}(e_{12}e_{31} - e_{31}e_{12}) = e_{23}$$

$$e_{31} \wedge e_{12} = \frac{1}{2}(e_{31}e_{12} - e_{12}e_{31}) = -e_{23}. \qquad (7.72)$$

These outer products are summarized in Table 7.17.

TABLE 7.17   Outer product of the 3D basis bivectors

Outer product	e_{23}	e_{31}	e_{12}
e_{23}		$-e_{12}$	e_{31}
e_{31}	e_{12}		$-e_{23}$
e_{12}	$-e_{31}$	e_{23}	

7.8.3 The geometric products

We see from the above analysis that the algebra for bivectors is identical to that for vectors, in that the symmetric component is a scalar, whilst the antisymmetric component is a bivector. Although this is true for three dimensions, it cannot be generalized to higher dimensions. For instance, in 4D there are six basis bivectors:

$$e_{12}\ e_{23}\ e_{34}\ e_{13}\ e_{14}\ e_{24} \qquad (7.73)$$

where some products create a scalar:

$$e_{12} \cdot e_{12} = -1 \qquad (7.74)$$

others, a bivector:

$$e_{12} \wedge e_{23} = e_{13} \qquad (7.75)$$

and others a grade-4 term:

$$e_{12} \wedge e_{34} = e_{1234}. \tag{7.76}$$

Table 7.18 combines Table 7.16 and 7.17 to create the geometric products for basis bivectors.

TABLE 7.18   Geometric product of the 3D basis bivectors

Geometric product	e_{23}	e_{31}	e_{12}
e_{23}	-1	$-e_{12}$	e_{31}
e_{31}	e_{12}	-1	$-e_{23}$
e_{12}	$-e_{31}$	e_{23}	-1

7.9 The bivector-trivector products

As mentioned above, the trivector is the highest dimensional object that exists in three dimensions, and determines the range of the outer product. Consequently, the products that do exist are the result of the inner product, and the geometric product simply reflects these results.

7.9.1 The inner product

Tables 7.19 and 7.20 summarize the inner products between the basis bivectors and unit basis trivector. These products also represent the dual relationships between vectors and bivectors in 3D. To maintain symmetry within the tables, the bivectors appear in the sequence e_{23}, e_{31}, e_{12}, and is continued in other tables.

TABLE 7.19   Inner product of the 3D basis bivectors

Inner product	I_{123}
e_{23}	$-e_1$
e_{31}	$-e_2$
e_{12}	$-e_3$

TABLE 7.20   Inner product of the 3D basis bivectors

Inner product	e_{23}	e_{31}	e_{12}
I_{123}	$-e_1$	$-e_2$	$-e_3$

Therefore, the pseudoscalar commutes with all bivectors in 3D. Observe how the product of a grade-3 pseudoscalar (I_{ijk}) with a grade-2 multivector (e_{ij}) is a grade $(3-2)$ multivector ($-e_k$).

The above can be generalized as follows:

In all spaces, I commutes with all even-grade multivectors in its algebra. And in odd dimensional spaces, I commutes with all multivectors. In even dimensional spaces, I anticommutes with all odd-grade multivectors.

These conditions are summarised in Table 7.21. It is also controlled by the following:

$$IA_r = (-1)^{r(n-1)}A_rI \tag{7.77}$$

where

$$I \text{ is the algebra's pseudoscalar}$$

$$A_r \text{ is a grade multivector}$$

$$n \text{ is the space dimension.}$$

For example, when $I = e_{123}, A_2 = e_{12}, n = 3$ then

$$e_{123}e_{12} = (-1)^{2(3-1)}e_{12}e_{123} \tag{7.78}$$

$$-e_3 = (-1)^4(-e_3) = -e_3. \tag{7.79}$$

TABLE 7.21   Commutation rules for the pseudoscalar and multivectors in different spaces

		multivectors	
		even grade	odd grade
spaces	even dimension	commute	anticommute
	odd dimension	commute	commute

7.10 The trivector-trivector product

The unit trivector is a pseudoscalar and squares to -1, which gives it its imaginary qualities. This is readily demonstrated as follows:

$$e_{123}e_{123} = e_{123123} = -e_{2332} = -1. \tag{7.80}$$

Once again, it is the inner product that gives rise to this result.

7.11 Product summary

Table 7.22 shows the inner products for the basis vectors, bivectors and trivector, Table 7.23 shows the corresponding outer products, and Table 7.24 combines them for the geometric product.

Table 7.22 Inner product of the 3D basis vectors, bivectors and trivector

Inner product	e_1	e_2	e_3	e_{23}	e_{31}	e_{12}	e_{123}
e_1	1				$-e_3$	e_2	e_{23}
e_2		1		e_3		$-e_1$	e_{31}
e_3			1	$-e_2$	e_1		e_{12}
e_{23}		$-e_3$	e_2	-1			$-e_1$
e_{31}	e_3		$-e_1$		-1		$-e_2$
e_{12}	$-e_2$	e_1				-1	$-e_3$
e_{123}	e_{23}	e_{31}	e_{12}	$-e_1$	$-e_2$	$-e_3$	-1

Table 7.23 Outer product of the 3D basis vectors, bivectors and trivector

Outer product	e_1	e_2	e_3	e_{23}	e_{31}	e_{12}	e_{123}
e_1		e_{12}	$-e_{31}$	e_{123}			
e_2	$-e_{12}$		e_{23}		e_{123}		
e_3	e_{31}	$-e_{23}$				e_{123}	
e_{23}	e_{123}				$-e_{12}$	e_{31}	
e_{31}		e_{123}		e_{12}		$-e_{23}$	
e_{12}			e_{123}	$-e_{31}$	e_{23}		
e_{123}							

Table 7.24 Geometric product of the 3D basis vectors, bivectors and trivector

Geometric product	e_1	e_2	e_3	e_{23}	e_{31}	e_{12}	e_{123}
e_1	1	e_{12}	$-e_{31}$	e_{123}	$-e_3$	e_2	e_{23}
e_2	$-e_{12}$	1	e_{23}	e_3	e_{123}	$-e_1$	e_{31}
e_3	e_{31}	$-e_{23}$	1	$-e_2$	e_1	e_{123}	e_{12}
e_{23}	e_{123}	$-e_3$	e_2	-1	$-e_{12}$	e_{31}	$-e_1$
e_{31}	e_3	e_{123}	$-e_1$	e_{12}	-1	$-e_{23}$	$-e_2$
e_{12}	$-e_2$	e_1	e_{123}	$-e_{31}$	e_{23}	-1	$-e_3$
e_{123}	e_{23}	e_{31}	e_{12}	$-e_1$	$-e_2$	$-e_3$	-1

TABLE 7.25   Classifying the geometric products

Geometric product	e_1	e_2	e_3	e_{23}	e_{31}	e_{12}	e_{123}
e_1	Scalars			Rotate the projected vector component 90°			D u a l s
e_2							
e_3	Bivectors			Trivectors			
e_{23}	Rotate the projected vector component −90°			Scalars			D u a l s
e_{31}							
e_{12}	Trivectors			Bivectors			
e_{123}	Duals			Duals			Scalar

Table 7.24 describes what is happening within each segment of Table 7.25 where:

- the vector-vector products creates scalars and bivectors
- the vector-bivector products rotate the projected vector component 90° and also create trivectors
- the vector-pseudoscalar products are duals
- the bivector-vector products rotate the projected vector component −90° and also create trivectors
- the bivector-bivector products creates scalars and bivectors
- the bivector-trivector products are duals
- the trivector-vector products are duals
- the trivector-bivector products are duals
- the trivector-trivector product is a scalar.

Tables 7.26, 7.27 and 7.28 summarize the actions of the inner, outer and geometric products for arbitrary vectors, bivectors and trivectors.

TABLE 7.26   Inner product summary

Inner product	b [vector]	B [bivector]	T [trivector]
a [vector]	$a \cdot b$ [scalar]	$a \cdot B$ [vector] rotated 90° in B	$a \cdot T$ [bivector]
A [bivector]	$A \cdot b$ [vector] rotated −90° in A	$-\|A\|\|B\|$ [scalar]	$A \cdot T$ [vector]
S [trivector]	$S \cdot b$ [bivector]	$S \cdot B$ [vector]	$-\|S\|\|T\|$ [scalar]

TABLE 7.27   Outer product summary

Outer product	b [vector]	B [bivector]	T [trivector]
a [vector]	$a \wedge b$ [bivector]	$a \wedge B$ [trivector]	
A [bivector]	$A \wedge b$ [trivector]		
S [trivector]			

TABLE 7.28   Geometric product summary

Geometric product	b [vector]	B [bivector]	T [trivector]
a [vector]	$a \cdot b + a \wedge b$ [multivector]	$a \cdot B + a \wedge B$ [multivector]	$a \cdot T$ [bivector]
A [bivector]	$A \cdot b + A \wedge b$ [multivector]	$-|A||B|$ [scalar]	$A \cdot T$ [vector]
S [trivector]	$S \cdot b$ [bivector]	$S \cdot B$ [vector]	$-|S||T|$ [scalar]

7.12 Geometric algebra and the vector product

The traditional vector product is a solid, reliable tool that creates a vector from two reference vectors and is perpendicular to both vectors. The right-hand rule determines the direction of the perpendicular vector. Unfortunately, it is ambiguous in spaces other than 3D, which is not a serious problem for most people, but it was a mistake and took vector analysis down a mathematical *cul de sac*.

Geometric algebra abandons the vector product in favour of Grassmann's outer product which starts its work in 2D and continues without limit. But the vector product's ghost is still lurking in the shadows in the form of bivector products. Let's explore its geometric alias.

7.13 The relationship between the vector product and outer product

We have already discovered that there is a very close relationship between the vector product and outer product, and just to recap:

Given two vectors

$$a = a_1 e_1 + a_2 e_2 + a_3 e_3$$
$$b = b_1 e_1 + b_2 e_2 + b_3 e_3 \qquad (7.81)$$

then

$$a \times b = (a_2 b_3 - a_3 b_2) e_1 + (a_3 b_1 - a_1 b_3) e_2 + (a_1 b_2 - a_2 b_1) e_3 \qquad (7.82)$$

and

$$a \wedge b = (a_2 b_3 - a_3 b_2)e_2 \wedge e_3 + (a_3 b_1 - a_1 b_3)e_3 \wedge e_1 + (a_1 b_2 - a_2 b_1)e_1 \wedge e_2 \qquad (7.83)$$

or

$$a \wedge b = (a_2 b_3 - a_3 b_2)e_{23} + (a_3 b_1 - a_1 b_3)e_{31} + (a_1 b_2 - a_2 b_1)e_{12}. \qquad (7.84)$$

If we multiply (7.68) by I_{123} we obtain

$$I_{123}(a \wedge b) = (a_2 b_3 - a_3 b_2)e_{123}e_{23} + (a_3 b_1 - a_1 b_3)e_{123}e_{31} + (a_1 b_2 - a_2 b_1)e_{123}e_{12}$$
$$= -(a_2 b_3 - a_3 b_2)e_1 - (a_3 b_1 - a_1 b_3)e_2 - (a_1 b_2 - a_2 b_1)e_3. \qquad (7.85)$$

Equation (7.69) is identical to (7.66) apart from its sign. Therefore, we can state:

$$a \times b = -I_{123}(a \wedge b). \qquad (7.86)$$

With such a simple relationship, geometric algebra has no need for the vector product as it can be created by multiplying the outer product by $-I_{123}$.

7.14 The relationship between quaternions and bivectors

When Hamilton had his amazing brain wave in 1843 and proposed the rules controlling quaternions, he must have thought that he had discovered a fundamentally new algebra. To a certain extent he had, but what he had really unearthed was the foundations of geometric algebra. The elegance and importance of quaternions seems to have made Hamilton oblivious to the parallel discoveries of Möbius and Grassmann.

Möbius had published his *barycentrische Calcul* in 1821 which described his ideas of directed line segments, centres of gravity, homogeneous coordinates, and signed areas and volumes. In 1832 Grassmann was using directed line segments and geometric products in his work on tides, and in 1844 he published his book *Die lineale Ausdehnungslehre*, that contained noncommutative vector products, *n*-dimensional spaces, vector differentiation, etc. Unfortunately, the book was not a success, nor widely read. When geometric algebra eventually emerged, it was realized that quaternions were closely related to bivectors, which we now examine.

Hamilton's rules for the imaginaries i, j and k are shown in Table 7.29, whilst Table 7.30 shows the rules for 3D bivector products.

TABLE 7.29 Hamilton's quaternion product rules

	i	j	k
i	-1	k	$-j$
j	$-k$	-1	i
k	j	$-i$	-1

Table 7.30   3D bivector product rules

	e_{23}	e_{31}	e_{12}
e_{23}	-1	$-e_{12}$	e_{31}
e_{31}	e_{12}	-1	$-e_{23}$
e_{12}	$-e_{31}$	e_{23}	-1

Although there is some agreement between the table entries, there is a sign reversal in the pink shaded entries. However, if we switch to a left-handed axial system the bivectors become e_{32}, e_{13}, e_{21} and their products are shown in Table 7.31.

Table 7.31   3D bivector product rules for a left-handed axial system

	e_{32}	e_{13}	e_{21}
e_{32}	-1	e_{21}	$-e_{13}$
e_{13}	$-e_{21}$	-1	e_{32}
e_{21}	e_{13}	$-e_{32}$	-1

If we now create the isomorphism

$$i \leftrightarrow e_{32} \quad j \leftrightarrow e_{13} \quad k \leftrightarrow e_{21} \tag{7.87}$$

there is a one-to-one correspondence between quaternions and a left-handed set of bivectors. Furthermore, Hamilton also introduced an extra rule where $ijk = -1$, and for a right-handed set of bivectors we have

$$e_{23}e_{31}e_{12} = +1 \tag{7.88}$$

whereas for a left-handed set of bivectors we have

$$e_{32}e_{13}e_{21} = -1 \tag{7.89}$$

which confirms the switch in axial systems.

7.15 The meet operation

The meet operation is used to identify the intersection of two blades and is often expressed as

$$L_1 \cup L_2 \equiv L_1^* \cdot L_2 \tag{7.90}$$

which implies that given two blades L_1 and L_2, their intersection $(L_1 \cup L_2)$ is computed by taking the inner product of the dual of L_1 with L_2, i.e. $L_1^* \cdot L_2$.

We can visualize this operation by considering two planes as shown in Figure 7.6.

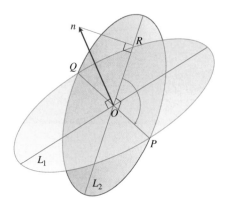

FIGURE 7.6.

The planes L_1 and L_2 intersect along PQ passing through O. If n is perpendicular to L_1 it is possible to project it on L_2 making OR, which, in turn, must be perpendicular to OP.

Geometric algebra provides us with the dual operation to compute a perpendicular vector, and the inner product of a vector and a bivector to rotate 90° in the plane of a bivector. For example, to compute the intersection of the blades e_{12} and e_{23} we proceed as follows:

$$e_{12} \cup e_{23} = e_{12}^* \cdot e_{23}$$

$$= (e_{123}e_{12}) \cdot e_{23}$$

$$= -e_3 \cdot e_{23}$$

$$= e_2. \tag{7.91}$$

The dual of e_{12} is $-e_3$, and when this rotated 90° in the e_{23} plane using the inner product we obtain e_2, as shown in Figure 7.7.

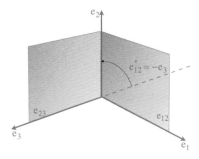

FIGURE 7.7.

Similarly,

$$e_{23} \cup e_{31} = e_{23}^* \cdot e_{31} = e_3$$

$$e_{31} \cup e_{12} = e_{31}^* \cdot e_{12} = e_1. \tag{7.92}$$

We explore other examples of the meet operation in later chapters.

7.16 Summary

In this chapter we have investigated every possible product that occurs between scalars, vectors, bivectors and trivectors. When a scalar is involved, the result is a simple scaling operation. All other combinations create a scalar, vector, bivector or a trivector. All of these products are shown in Table 7.26 which also identifies those products that give rise to dual relationships.

So far, we have only considered two- and three-dimensional space and geometric algebra does not recognise any limits to the number of dimensions a space may possess. Therefore, in spaces of higher dimensions, the same approach can be used to identify products involving higher order objects.

We have also shown that there is no need for the traditional vector (cross) product, as it is a dual of the outer product of two vectors. Similarly, quaternions are not something special in their own right, but are nothing more than a left-handed system of bivectors. When we show how geometric algebra handles rotations, the mystery of quaternions evaporates and, like the vector product, can be quietly put to one side.

8 Reflections and Rotations

8.1 Introduction

In this chapter we investigate how geometric algebra introduces totally new ways to calculate reflections and rotations. Although the operations take place about the origin, standard vectorial techniques enable them to be translated to other positions in space. We begin with reflections as they provide a cunning way to implement rotations. The first type of reflection is relative to a line and the second type is relative to a mirror plane.

8.2 Reflections using normal vectors

Let's start with a problem often encountered in computer games and animation where we need to compute the reflected ray from an incident ray. We start with the 2D scenario shown in Figure 8.1 where vector a subtends an angle α with the unit vector $\hat{m}$. The objective is to compute the reflected vector a' which also subtends an angle α with $\hat{m}$.

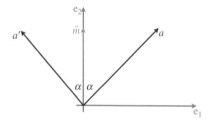

FIGURE 8.1.

If

$$a = a_1 e_1 + a_2 e_2 \tag{8.1}$$

J. Vince, *Geometric Algebra: An Algebraic System for Computer Games and Animation*,
© Springer-Verlag London Limited 2009

then, by inspection

$$a' = -a_1 e_1 + a_2 e_2. \tag{8.2}$$

Basically, we need to preserve the sign of the component of a parallel to $\hat{m}$ but flip the sign of the component of a perpendicular to $\hat{m}$.

The solution to this problem is found in the geometric product of two vectors. For example, Figure 8.2 shows a unit vector $\hat{m}$ perpendicular to a plane P, where vector a subtends an angle α with $\hat{m}$; so, too, does its reflection a'.

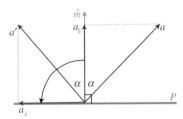

FIGURE 8.2.

Taking the inner product $\hat{m} \cdot a$ and using this to scale $\hat{m}$, we obtain $a_\parallel$:

$$a_\parallel = (\hat{m} \cdot a)\hat{m}. \tag{8.3}$$

Similarly, taking the outer product $a \wedge \hat{m}$ and multiplying this by $\hat{m}$, we obtain $a_\perp$:

$$a_\perp = \hat{m}(a \wedge \hat{m}) = (\hat{m} \wedge a)\hat{m}. \tag{8.4}$$

Multiplying $\hat{m}$ by the bivector $a \wedge \hat{m}$ rotates it 90° and creates $a_\perp$. See (6.56).

From Figure 8.2

$$a' = a_\parallel + a_\perp \tag{8.5}$$

therefore,

$$a' = (\hat{m} \cdot a)\hat{m} + (\hat{m} \wedge a)\hat{m}$$
$$= (\hat{m} \cdot a + \hat{m} \wedge a)\hat{m}$$
$$a' = \hat{m}a\hat{m} \tag{8.6}$$

which is rather elegant!

Although the above scenario is two-dimensional, the concept generalizes to higher dimensions, and to illustrate this, consider the 3D scenario in Figure 8.3 where

$$a = e_1 + e_2 - e_3 \tag{8.7}$$

and

$$\hat{m} = e_2. \tag{8.8}$$

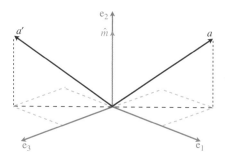

Figure 8.3.

Therefore, using (8.6)

$$a' = \hat{m}a\hat{m}$$

$$= e_2(e_1 + e_2 - e_3)e_2$$

$$= e_{212} + e_{222} - e_{232}$$

$$a' = -e_1 + e_2 + e_3 \tag{8.9}$$

which is correct.

Instead of using a normal vector to represent the orientation of a surface, we can use a unit bivector $\hat{B}$ as shown in Figure 8.4. The bivector's normal is represented by its dual $\hat{B}^*$, and even though $-\hat{B}^*$ represents $\hat{m}$ in the above, it makes no difference to our calculations. Therefore, (8.6) becomes

$$a' = \hat{B}^*a\hat{B}^*. \tag{8.10}$$

Figure 8.4 shows a bivector equivalent of Figure 8.3 where

$$\hat{B} = e_{31} \tag{8.11}$$

$$a = e_1 + e_2 - e_3 \tag{8.12}$$

therefore,

$$a' = \hat{B}^*a\hat{B}^*$$

$$= e_{123}e_{31}(e_1 + e_2 - e_3)e_{123}e_{31}$$

$$= e_{12331}(e_1 + e_2 - e_3)e_{12331}$$

$$= -e_2(e_1 + e_2 - e_3)(-e_2)$$

$$= e_{212} + e_{222} - e_{232}$$

$$a' = -e_1 + e_2 + e_3 \tag{8.13}$$

and is correct.

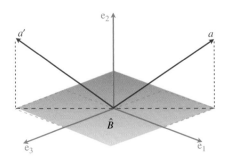

Figure 8.4.

In 2D we can reflect bivectors about a vector and in 3D we can reflect bivectors and trivectors. For instance, two 2D vectors a and b have reflections a' and b' respectively about $\hat{m}$ where

$$a' = \hat{m}a\hat{m}$$
$$b' = \hat{m}b\hat{m}. \tag{8.14}$$

If

$$B = a \wedge b \tag{8.15}$$

then

$$B' = a' \wedge b'$$
$$= (\hat{m}a\hat{m}) \wedge (\hat{m}b\hat{m}). \tag{8.16}$$

Using the identity

$$B' = \frac{1}{2}(a'b' - b'a') \tag{8.17}$$

then

$$B' = \frac{1}{2}[(\hat{m}a\hat{m})(\hat{m}b\hat{m}) - (\hat{m}b\hat{m})(\hat{m}a\hat{m})]$$
$$= \frac{1}{2}(\hat{m}a\hat{m}\hat{m}b\hat{m} - \hat{m}b\hat{m}\hat{m}a\hat{m})$$
$$= \frac{1}{2}(\hat{m}ab\hat{m} - \hat{m}ba\hat{m})$$
$$= \frac{1}{2}\hat{m}(ab - ba)\hat{m}$$
$$B' = \hat{m}B\hat{m}. \tag{8.18}$$

For example, Figure 8.5 shows three vectors a, b and $\hat{m}$ where

$$a = e_1$$
$$b = e_1 + e_2$$
$$\hat{m} = e_2. \tag{8.19}$$

Therefore,

$$B = a \wedge b$$
$$= e_1 \wedge (e_1 + e_2)$$
$$B = e_{12} \tag{8.20}$$

and

$$a' = \hat{m}a\hat{m}$$
$$= e_2 e_1 e_2$$
$$a' = -e_1$$

$$b' = \hat{m}b\hat{m}$$
$$= e_2(e_1 + e_2)e_2$$
$$b' = -e_1 + e_2$$

$$B' = a' \wedge b'$$
$$= -e_1 \wedge (-e_1 + e_2)$$
$$B' = -e_{12}. \tag{8.21}$$

Using (8.18) we have

$$B' = \hat{m}B\hat{m}$$
$$= e_2 e_{12} e_2$$
$$B' = -e_{12} \tag{8.22}$$

which confirms (8.21).

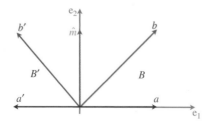

FIGURE 8.5.

Now let's consider the unit basis trivector, which is also the pseudoscalar. We define the trivector as follows:

$$T = e_1 \wedge e_2 \wedge e_3$$

$$T = e_1 e_2 e_3. \tag{8.23}$$

Each unit basis vector has its own reflection:

$$e_1' = \hat{m} e_1 \hat{m}$$

$$e_2' = \hat{m} e_2 \hat{m}$$

$$e_3' = \hat{m} e_3 \hat{m} \tag{8.24}$$

therefore, the reflected trivector T' is

$$T' = (\hat{m} e_1 \hat{m}) \wedge (\hat{m} e_2 \hat{m}) \wedge (\hat{m} e_3 \hat{m})$$

$$= \hat{m} e_1 \hat{m} \hat{m} e_2 \hat{m} \hat{m} e_3 \hat{m}$$

$$= \hat{m} e_1 e_2 e_3 \hat{m}$$

$$T' = \hat{m} T \hat{m} \tag{8.25}$$

which continues the pattern for vectors and bivectors.

8.3 Reflecting a vector in a mirror

Next, we consider reflecting a vector in a planar mirror. The orientation of the mirror is determined by the unit normal vector $\hat{m}$ as shown in Figure 8.6, which also shows vector a and its reflection a'.

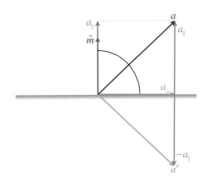

Figure 8.6.

Once again,

$$a_\parallel = (\hat{m} \cdot a)\hat{m} \tag{8.26}$$

but this time

$$a_\perp = (a \wedge \hat{m})\hat{m}$$

$$a_\perp = -(\hat{m} \wedge a)\hat{m}. \tag{8.27}$$

Therefore,

$$a' = -a_\parallel + a_\perp$$

$$= -(\hat{m} \cdot a)\hat{m} - (\hat{m} \wedge a)\hat{m}$$

$$a' = -\hat{m}a\hat{m}. \tag{8.28}$$

Note that we could have arrived at this result simply by observing that a' is the negative of the reflection about $\hat{m}$. Furthermore, if the mirror is represented by the bivector $\hat{B}$ then

$$a' = -\hat{B}^* a \hat{B}^*. \tag{8.29}$$

This equation is used for reflecting rays from surfaces.

There is no need to illustrate the action of (8.29) as it should be self evident.

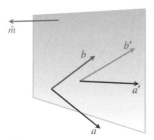

FIGURE 8.7.

8.4 Reflecting a bivector in a mirror

We can compute the reflection of a bivector by reflecting the individual vectors forming the bivector and reconstructing the reflected bivector. For example, if

$$B = a \wedge b \tag{8.30}$$

then the reflected bivector is

$$B' = a' \wedge b' \tag{8.31}$$

where a' and b' are the reflections of a and b respectively. Consequently, if $\hat{m}$ represents the mirror's surface normal, as shown in Figure 8.7, then

$$a' = -\hat{m}a\hat{m}$$

$$b' = -\hat{m}b\hat{m} \tag{8.32}$$

and

$$B' = (-\hat{m}a\hat{m}) \wedge (-\hat{m}b\hat{m})$$
$$= (\hat{m}a\hat{m}) \wedge (\hat{m}b\hat{m}). \tag{8.33}$$

Using the identity

$$B' = \tfrac{1}{2}(a'b' - b'a')$$
$$= \tfrac{1}{2}(\hat{m}a\hat{m}\hat{m}b\hat{m} - \hat{m}b\hat{m}\hat{m}a\hat{m})$$
$$= \tfrac{1}{2}(\hat{m}ab\hat{m} - \hat{m}ba\hat{m})$$
$$= \tfrac{1}{2}\hat{m}(ab - ba)\hat{m}$$
$$B' = \hat{m}B\hat{m}. \tag{8.34}$$

This time there is no minus sign. And to illustrate (8.34) consider the following:

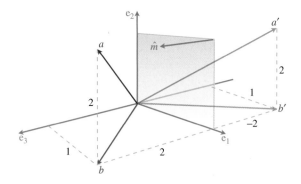

FIGURE 8.8.

Figure 8.8 shows a scenario where a mirror with surface normal $\hat{m}$ exists in the e_{12} plane. The original vectors are

$$a = e_1 + 2e_2 + 2e_3$$
$$b = e_1 + 2e_3$$
$$\hat{m} = e_3 \tag{8.35}$$

and

$$B = a \wedge b$$
$$= (e_1 + 2e_2 + 2e_3) \wedge (e_1 + 2e_3)$$
$$= 2e_{13} + 2e_{21} + 4e_{23} + 2e_{31}$$
$$B = -2e_{12} + 4e_{23} \tag{8.36}$$

then

$$B' = \hat{m}B\hat{m}$$

$$= e_3(-2e_{12} + 4e_{23})\,e_3$$

$$= -2e_{3123} + 4e_{3233}$$

$$B' = -2e_{12} - 4e_{23}. \tag{8.37}$$

We can confirm this result by computing $a' \wedge b'$:

$$a' \wedge b' = (e_1 + 2e_2 - 2e_3) \wedge (e_1 - 2e_3)$$

$$= -2e_{13} + 2e_{21} - 4e_{23} - 2e_{31}$$

$$= -2e_{12} - 4e_{23}$$

$$a' \wedge b' = B'. \tag{8.38}$$

8.5 Reflecting a trivector in a mirror

The axial convention employed in this book is right-handed as it is widely used in computer games and animation, consequently geometric algebra's trivector is also right-handed. We know that when we hold up our right hand in front of a mirror, we observe that its reflection is left-handed. This suggests that a trivector's reflection is also left-handed, which must introduce a sign change. This is demonstrated as follows:

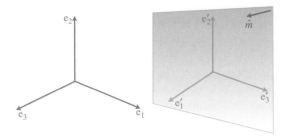

FIGURE 8.9.

Figure 8.9 shows a right-handed set of axes $[e_1 e_2 e_3]$ with its reflection $[e'_1 e'_2 e'_3]$. The mirror's normal is $\hat{m}$.

Starting with a unit trivector, which is also the pseudoscalar, we have:

$$T = e_1 \wedge e_2 \wedge e_3$$

$$T = e_1 e_2 e_3. \tag{8.39}$$

Each basis vector has its own reflection:

$$e_1' = -\hat{m}e_1\hat{m}$$
$$e_2' = -\hat{m}e_2\hat{m}$$
$$e_3' = -\hat{m}e_3\hat{m} \qquad (8.40)$$

therefore, the reflected trivector T' is

$$T' = (-\hat{m}e_1\hat{m}) \wedge (-\hat{m}e_2\hat{m}) \wedge (-\hat{m}e_3\hat{m})$$
$$= -\hat{m}e_1\hat{m}\hat{m}e_2\hat{m}\hat{m}e_3\hat{m}$$
$$= -\hat{m}e_1e_2e_3\hat{m}$$
$$T' = -\hat{m}T\hat{m} \qquad (8.41)$$

and as anticipated, a sign reversal has occurred.

8.6 Reflecting scalars

Scalars are a grade 0 element and have no physical extent and therefore are invariant under reflections.

To summarize, we have the following rules for reflections:

Reflecting about a line

$$
\begin{array}{ll}
\textit{scalars} & \text{invariant} \\
\textit{vectors} & v' = \hat{m}v\hat{m} \\
\textit{bivectors} & B' = \hat{m}B\hat{m} \\
\textit{trivectors} & T' = \hat{m}T\hat{m}.
\end{array} \qquad (8.42)
$$

Reflecting in a mirror

$$
\begin{array}{ll}
\textit{scalars} & \text{invariant} \\
\textit{vectors} & v' = -\hat{m}v\hat{m} \\
\textit{bivectors} & B' = \hat{m}B\hat{m} \\
\textit{trivectors} & T' = -\hat{m}T\hat{m}.
\end{array} \qquad (8.43)
$$

8.7 Rotations by reflections

A mirror creates a virtual space behind its reflecting surface where we visualize reflections of physical points. If we countenance this virtual space with some sort of physical reality, as shown in Figure 8.10, points P_1 and P_2 have reflections P_1' and P_2' respectively.

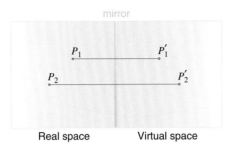

mirror

P_1 ———— P'_1

P_2 ———— P'_2

Real space   Virtual space

FIGURE 8.10.

Hopefully, it is obvious that, in general, a rotation transform cannot be used to compute these reflections. However, it just so happens that two successive reflections give rise to a rotation, which provides geometric algebra with a rather elegant alternative to the traditional rotation transform.

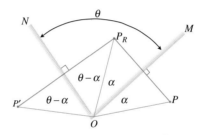

FIGURE 8.11.

To understand this process, consider the two mirrors M and N separated by an angle θ, in Figure 8.11. The figure shows a plan view of a 3D scene. The point P is in front of mirror M and subtends an angle α, and its reflection P_R exists in the virtual space behind M and also subtends an angle α with the mirror.

The angle between P_R and N must be $\theta - \alpha$, and its reflection P' must also lie $\theta - \alpha$ behind N. By inspection, the angle between P and the double reflection P' is 2θ.

If we apply this double reflection transform to a collection of points, they are effectively all rotated 2θ about the origin where the mirrors intersect. The only slight drawback is that the angle of rotation is twice the angle between the mirrors.

Instead of using points, let's employ position vectors and substitute normal unit vectors for the mirror's orientations. For example, Figure 8.12 shows the same mirrors with unit normal vectors $\hat{m}$ and $\hat{n}$. After two successive reflections, P becomes P', and using (8.28) we compute the reflection as follows:

$$p_R = -\hat{m}p\hat{m}$$
$$p' = -\hat{n}p_R\hat{n}$$
$$p' = \hat{n}\hat{m}p\hat{m}\hat{n} \tag{8.44}$$

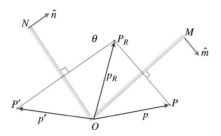

FIGURE 8.12.

which is also rather elegant and compact. However, we must remember that P is rotated twice the angle separating the mirrors, and the rotation is relative to the origin. Let's demonstrate this technique with a simple example.

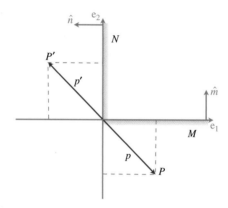

FIGURE 8.13.

Figure 8.13 shows two mirrors M and N with unit normal vectors

$$\hat{m} = e_2$$

$$\hat{n} = -e_1$$

$$p = e_1 - e_2. \tag{8.45}$$

As the mirrors are separated by 90° the point P is rotated 180°:

$$p' = \hat{n}\hat{m}p\hat{m}\hat{n}$$

$$= -e_1 e_2 (e_1 - e_2) e_2 (-e_1)$$

$$= e_{12121} - e_{12221}$$

$$p' = -e_1 + e_2 \tag{8.46}$$

which is correct.

Although the reflecting qualities of mirrors provide a useful model to illustrate how two reflections create a rotation, we can abandon this physical model and adopt another provided by geometric algebra.

We see from (8.44) that it is $\hat{n}$ and $\hat{m}$ that control the angle and plane of rotation. Furthermore, the geometric products $\hat{m}\hat{n}$ and $\hat{n}\hat{m}$ encode $\cos\theta$ within the inner product and $\sin\theta$ within the outer product, therefore, the bivector $\hat{m}\wedge\hat{n}$ is sufficient to specify the plane and angle of rotation. For example, consider the scenario shown in Figure 8.14.

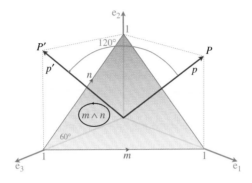

FIGURE 8.14.

The two vectors m and n are used to define the bivector $m\wedge n$, and as these create an equilateral triangle, the interior angle is 60°.

Using unit vectors $\hat{m}$ and $\hat{n}$

$$\hat{n}\hat{m}p\hat{m}\hat{n} \tag{8.47}$$

rotates p 120° anticlockwise about the plane $\hat{m}\wedge\hat{n}$.

Given

$$p = e_1 + e_2$$

$$m = e_1 - e_3$$

$$\hat{m} = \frac{1}{\sqrt{2}}(e_1 - e_3)$$

$$n = e_2 - e_3$$

$$\hat{n} = \frac{1}{\sqrt{2}}(e_2 - e_3) \tag{8.48}$$

therefore,

$$\hat{n}\hat{m} = \tfrac{1}{2}(e_2 - e_3)(e_1 - e_3)$$

$$= \tfrac{1}{2}(e_{21} - e_{23} - e_{31} + 1)$$

$$\hat{m}\hat{n} = \tfrac{1}{2}(e_1 - e_3)(e_2 - e_3)$$

$$= \tfrac{1}{2}(e_{12} - e_{13} - e_{32} + 1) \tag{8.49}$$

and

$$p' = \hat{n}\hat{m}p\hat{m}\hat{n}$$

$$= \tfrac{1}{4}(e_{21} - e_{23} - e_{31} + 1)(e_1 + e_2)(e_{12} - e_{13} - e_{32} + 1)$$

$$= \tfrac{1}{4}(e_2 - e_1 - e_{231} + e_3 - e_3 - e_{312} + e_1 + e_2)(e_{12} - e_{13} - e_{32} + 1)$$

$$= \tfrac{1}{4}(2e_2 - 2e_{123})(e_{12} - e_{13} - e_{32} + 1)$$

$$= \tfrac{1}{4}(-2e_1 - 2e_{213} + 2e_3 + 2e_2 - 2e_{12312} + 2e_{12313} + 2e_{12332} - 2e_{123})$$

$$= \tfrac{1}{2}(-e_1 + e_{123} + e_3 + e_2 + e_3 + e_2 + e_1 - e_{123})$$

$$= \tfrac{1}{2}(2e_2 + 2e_3)$$

$$p' = e_2 + e_3 \qquad\qquad (8.50)$$

which is correct.

The above analysis may seem rather verbose, but such operations are normally carried out by software.

Although the expression $\hat{n}\hat{m}p\hat{m}\hat{n}$ may appear rather strange for a rotation transform, it is closely related to the traditional rotation matrix. For example, consider the 2D scenario shown in Figure 8.15 where vectors $\hat{m}$ and $\hat{n}$ are separated by an angle θ. The vector p is rotated through an angle 2θ to p' using

$$p' = \hat{n}\hat{m}p\hat{m}\hat{n}. \qquad\qquad (8.51)$$

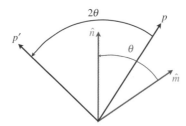

FIGURE 8.15.

We now expand (8.51) and show that it is nothing more than a rotation matrix in disguise!

Let

$$\hat{m} = m_1 e_1 + m_2 e_2$$

$$\hat{n} = n_1 e_1 + n_2 e_2$$

$$p = p_1 e_1 + p_2 e_2 \qquad\qquad (8.52)$$

and θ is the angle between $\hat{m}$ and $\hat{n}$.

Therefore,

$$\hat{m}\hat{n} = \hat{m} \cdot \hat{n} + \hat{m} \wedge \hat{n} \qquad\qquad (8.53)$$

and

$$\hat{n}\hat{m} = \hat{m} \cdot \hat{n} + \hat{n} \wedge \hat{m} \tag{8.54}$$

where

$$\hat{m} \cdot \hat{n} = \cos\theta$$
$$\hat{m} \wedge \hat{n} = \sin\theta e_{12}$$
$$\hat{n} \wedge \hat{m} = -\sin\theta e_{12}. \tag{8.55}$$

Then

$$
\begin{aligned}
p' &= \hat{n}\hat{m}p\hat{m}\hat{n} \\
&= (\cos\theta - \sin\theta e_{12})(p_1 e_1 + p_2 e_2)(\cos\theta + \sin\theta e_{12}) \\
&= (p_1\cos\theta e_1 + p_2\cos\theta e_2 + p_1\sin\theta e_2 - p_2\sin\theta e_1)(\cos\theta + \sin\theta e_{12}) \\
&= ((p_1\cos\theta - p_2\sin\theta)e_1 + (p_1\sin\theta + p_2\cos\theta)e_2)(\cos\theta + \sin\theta e_{12}) \\
&= (p_1(\cos^2\theta - \sin^2\theta) - p_2 2\cos\theta\sin\theta)e_1 + (p_1(2\cos\theta\sin\theta) + p_2(\cos^2\theta - \sin^2\theta))e_2 \\
&= (p_1\cos 2\theta - p_2\sin 2\theta)e_1 + (p_1\sin 2\theta + p_2\cos 2\theta)e_2 \tag{8.56}
\end{aligned}
$$

or in matrix form

$$
\begin{bmatrix} p'_1 \\ p'_2 \end{bmatrix} = \begin{bmatrix} \cos 2\theta & -\sin 2\theta \\ \sin 2\theta & \cos 2\theta \end{bmatrix} \begin{bmatrix} p_1 \\ p_2 \end{bmatrix} \tag{8.57}
$$

which is the rotation matrix for 2θ!

8.8 Rotors

One of the powerful features of geometric algebra is the ease with which vectors can be rotated. For example, the inner product of a vector and a bivector rotates the vector 90° about the bivector's normal vector. And although this seems a rather obscure feature, it does have its uses.

We have also just discovered that a vector is rotated about a bivector's normal vector by an angle twice that of the angle associated with the bivector:

$$p' = \hat{n}\hat{m}p\hat{m}\hat{n}. \tag{8.58}$$

Which is fine when $\hat{n}$ and $\hat{m}$ have simple orientations, but presents a problem if we want to rotate vector p some arbitrary angle about an arbitrary axis.

Quaternions are the solution to rotations of arbitrary angles about arbitrary axes, and although it may not be immediately obvious, we have already started to discover geometric algebra's equivalent.

We begin by substituting R for $\hat{n}\hat{m}$ and $\tilde{R} = \hat{m}\hat{n}$, therefore,

$$p' = Rp\tilde{R} \tag{8.59}$$

where R and $\tilde{R}$ are called *rotors*.

We now unpack a rotor in terms of its angle and bivector as follows:

The bivector defining the plane is $\hat{m} \wedge \hat{n}$ and θ is the angle between the vectors. Let

$$R = \hat{n}\hat{m} \tag{8.60}$$

and

$$\tilde{R} = \hat{m}\hat{n} \tag{8.61}$$

where

$$\hat{n}\hat{m} = \hat{n} \cdot \hat{m} - \hat{m} \wedge \hat{n} \tag{8.62}$$

and

$$\hat{m}\hat{n} = \hat{n} \cdot \hat{m} + \hat{m} \wedge \hat{n}. \tag{8.63}$$

But

$$\hat{n} \cdot \hat{m} = \cos\theta \tag{8.64}$$

and

$$\hat{m} \wedge \hat{n} = \hat{B}\sin\theta \tag{8.65}$$

therefore,

$$R = \cos\theta - \hat{B}\sin\theta$$
$$\tilde{R} = \cos\theta + \hat{B}\sin\theta. \tag{8.66}$$

We now have an equation that rotates a vector p through an angle 2θ about an axis defined by $\hat{B}$:

$$p' = Rp\tilde{R} \tag{8.67}$$

or

$$p' = (\cos\theta - \hat{B}\sin\theta)p(\cos\theta + \hat{B}\sin\theta). \tag{8.68}$$

We can also write (8.68) such that it identifies the real angle of rotation α:

$$p' = \left(\cos(\alpha/2) - \hat{B}\sin(\alpha/2)\right) p \left(\cos(\alpha/2) + \hat{B}\sin(\alpha/2)\right). \tag{8.69}$$

There is still a small problem with (8.69) in that it references a bivector $\hat{B}$, rather than an axis.

However, this is easily rectified by selecting a vector a and creating its dual by forming the product

$$B = e_{123}a = Ia. \tag{8.70}$$

For example, if

$$a = e_2 \tag{8.71}$$

then

$$B = e_{123}e_2 = e_{31} \tag{8.72}$$

which provides the correct orientation of B with respect to a. The bivector B must then be normalized to produce $\hat{B}$.

So, the expression to rotate the vector p, angle α about an axis a becomes:

$$p' = (\cos(\alpha/2) - Ia\sin(\alpha/2)) p (\cos(\alpha/2) + Ia\sin(\alpha/2)). \tag{8.73}$$

Care must be taken with the orientation of the bivector $\hat{B}$ and the vector a, for if there is a sign reversal, the direction of rotation is reversed.

Let's demonstrate how (8.73) works with two examples.

Example 1 Figure 8.16 shows a scenario where vector p is rotated 90° about e_2, where

$$\alpha = 90°$$

$$a = e_2$$

$$p = e_1 + e_2$$

$$\hat{B} = Ia = e_{123}e_2 = e_{31}. \tag{8.74}$$

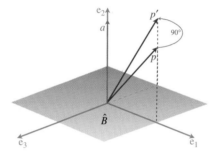

FIGURE 8.16.

$$p' = (\cos 45° - e_{31}\sin 45°)\,(e_1 + e_2)\,(\cos 45° + e_{31}\sin 45°)$$

$$= \left(\frac{\sqrt{2}}{2} - \frac{\sqrt{2}}{2}e_{31}\right)(e_1 + e_2)\left(\frac{\sqrt{2}}{2} + \frac{\sqrt{2}}{2}e_{31}\right)$$

$$= \left(\frac{\sqrt{2}}{2}e_1 + \frac{\sqrt{2}}{2}e_2 - \frac{\sqrt{2}}{2}e_3 - \frac{\sqrt{2}}{2}e_{312}\right)\left(\frac{\sqrt{2}}{2} + \frac{\sqrt{2}}{2}e_{31}\right)$$

$$= \frac{e_1}{2} - \frac{e_3}{2} + \frac{e_2}{2} + \frac{e_{231}}{2} - \frac{e_3}{2} - \frac{e_1}{2} - \frac{e_{312}}{2} - \frac{e_{31231}}{2}$$

$$p' = e_2 - e_3 \tag{8.75}$$

which is correct.

Observe what happens when the bivector's sign is reversed to $-e_{31}$:

$$p' = (\cos 45° + e_{31}\sin 45°)\,(e_1 + e_2)\,(\cos 45° - e_{31}\sin 45°)$$

$$= \left(\frac{\sqrt{2}}{2} + \frac{\sqrt{2}}{2}e_{31}\right)(e_1 + e_2)\left(\frac{\sqrt{2}}{2} - \frac{\sqrt{2}}{2}e_{31}\right)$$

$$= \left(\frac{\sqrt{2}}{2}e_1 + \frac{\sqrt{2}}{2}e_2 + \frac{\sqrt{2}}{2}e_3 + \frac{\sqrt{2}}{2}e_{312}\right)\left(\frac{\sqrt{2}}{2} - \frac{\sqrt{2}}{2}e_{31}\right)$$

$$= \frac{e_1}{2} + \frac{e_3}{2} + \frac{e_2}{2} + \frac{e_{231}}{2} + \frac{e_3}{2} + \frac{e_1}{2} + \frac{e_{312}}{2} - \frac{e_{31231}}{2}$$

$$p' = e_2 + e_3 \tag{8.76}$$

the rotation is clockwise about e_2.

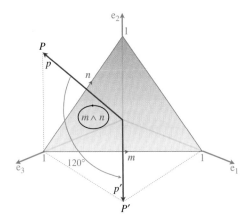

FIGURE 8.17.

Example 2 Figure 8.17 shows a scenario where vector p is rotated $120°$ about B, where

$$m = e_1 - e_3$$

$$n = e_2 - e_3$$

$$\alpha = 120°$$

$$p = e_2 + e_3$$

$$B = m \wedge n$$

$$= (e_1 - e_3)(e_2 + e_3)$$

$$B = e_{12} + e_{31} + e_{23}. \tag{8.77}$$

Next, we normalize B to $\hat{B}$:

$$\hat{B} = \frac{1}{\sqrt{3}}(e_{12} + e_{23} + e_{31}) \tag{8.78}$$

therefore,

$$p' = (\cos 60° - \hat{B}\sin 60°)\, p\, (\cos 60° + \hat{B}\sin 60°)$$

$$= \left(\cos 60° - \frac{1}{\sqrt{3}}(e_{12}+e_{23}+e_{31})\sin 60°\right)(e_2+e_3)\left(\cos 60° + \frac{1}{\sqrt{3}}(e_{12}+e_{23}+e_{31})\sin 60°\right)$$

$$= \left(\frac{1}{2} - \frac{e_{12}}{2} - \frac{e_{23}}{2} - \frac{e_{31}}{2}\right)(e_2+e_3)\left(\frac{1}{2} + \frac{e_{12}}{2} + \frac{e_{23}}{2} + \frac{e_{31}}{2}\right)$$

$$= \left(\frac{e_2}{2} + \frac{e_3}{2} - \frac{e_1}{e_2} - \frac{e_{123}}{2} + \frac{e_3}{2} - \frac{e_2}{2} - \frac{e_{312}}{2} + \frac{e_1}{2}\right)\left(\frac{1}{2} + \frac{e_{12}}{2} + \frac{e_{23}}{2} + \frac{e_{31}}{2}\right)$$

$$= (e_3 - e_{123})\left(\frac{1}{2} + \frac{e_{12}}{2} + \frac{e_{23}}{2} + \frac{e_{31}}{2}\right)$$

$$= \frac{e_3}{2} + \frac{e_{312}}{2} - \frac{e_2}{2} + \frac{e_1}{2} - \frac{e_{123}}{2} - \frac{e_{12312}}{2} - \frac{e_{12323}}{2} - \frac{e_{12331}}{2}$$

$$= \frac{e_3}{2} - \frac{e_2}{2} + \frac{e_1}{2} + \frac{e_3}{2} + \frac{e_1}{2} + \frac{e_2}{2}$$

$$p' = e_1 + e_3 \tag{8.79}$$

which is correct.

8.9 Rotating bivectors

We have already demonstrated that a bivector C is reflected using

$$C_R = \hat{m}C\hat{m} \tag{8.80}$$

and there is nothing stopping us from subjecting it to a double reflection and rotating it using

$$C' = \hat{n}\hat{m}C\hat{m}\hat{n} \tag{8.81}$$

or

$$C' = (\cos(\alpha/2) - \hat{B}\sin(\alpha/2))C(\cos(\alpha/2) + \hat{B}(\alpha/2)). \tag{8.82}$$

For example, consider the scenario in Figure 8.18 where

$$\hat{B} = e_{31}$$

$$C = e_{12}$$

$$\alpha = 90° \tag{8.83}$$

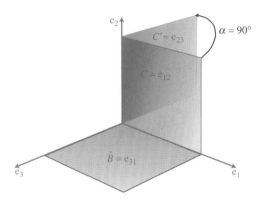

FIGURE 8.18.

then

$$C' = \left(\frac{\sqrt{2}}{2} - e_{31}\frac{\sqrt{2}}{2}\right) e_{12} \left(\frac{\sqrt{2}}{2} + e_{31}\frac{\sqrt{2}}{2}\right)$$

$$= \left(\frac{\sqrt{2}}{2}e_{12} - \frac{\sqrt{2}}{2}e_{312}\right)\left(\frac{\sqrt{2}}{2} + e_{31}\frac{\sqrt{2}}{2}\right)$$

$$= \frac{e_{12}}{2} + \frac{e_{1231}}{2} - \frac{e_{32}}{2} - \frac{e_{3231}}{2}$$

$$= \frac{e_{12}}{2} + \frac{e_{23}}{2} + \frac{e_{23}}{2} - \frac{e_{12}}{2}$$

$$C' = e_{23} \tag{8.84}$$

which is correct.

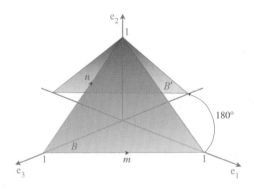

FIGURE 8.19.

Next, consider the scenario shown in Figure 8.19 where we see the bivector B rotated $180°$ about e_2.

The bivector

$$B = m \wedge n \tag{8.85}$$

has equal components in the three basis bivector planes:

$$m = e_1 - e_3$$

$$n = e_2 - e_3$$

$$B = m \wedge n$$

$$= (e_1 - e_3)(e_2 - e_3)$$

$$= e_{12} - e_{13} - e_{32}$$

$$B = e_{12} + e_{23} + e_{31}. \tag{8.86}$$

As the bivector is being rotated about an axis perpendicular to e_{31}, the e_{31} component of B remains unchanged. The other two bivector components will flip signs, thus we expect

$$B' = -e_{12} - e_{23} + e_{31}. \tag{8.87}$$

Let's rotate B 180° about e_2 , i.e. e_{31}:

$$B' = (\cos 90° - e_{31} \sin 90°)(e_{12} + e_{23} + e_{31})(\cos 90° + e_{31} \sin 90°)$$

$$= -e_{31}(e_{12} + e_{23} + e_{31})e_{31}$$

$$= -e_{311231} - e_{312331} - e_{313131}$$

$$B' = -e_{12} - e_{23} + e_{31} \tag{8.88}$$

which is as predicted.

8.10 Rotating trivectors

We have previously demonstrated that a trivector is reflected using

$$T_R = \hat{m} T \hat{m} \tag{8.89}$$

and as with bivectors, there is nothing preventing us from subjecting it to a double reflection and rotating it using

$$T' = -\hat{n}(-\hat{m} T \hat{m})\hat{n}$$

$$T' = \hat{n}\hat{m} T \hat{m}\hat{n} \tag{8.90}$$

or

$$T' = \left(\cos(\alpha/2) - \hat{B} \sin(\alpha/2) \sin(\alpha/2) \right) T \left(\cos(\alpha/2) + \hat{B} \sin(\alpha/2) \sin(\alpha/2) \right) \tag{8.91}$$

or

$$T' = (\cos(\alpha/2) - Ia \sin(\alpha/2) \sin(\alpha/2)) \, T \, (\cos(\alpha/2) + Ia \sin(\alpha/2) \sin(\alpha/2)) \,. \tag{8.92}$$

For example, if

$$a = e_1$$

$$b = e_2$$

$$c = e_3$$

$$T = a \wedge b \wedge c$$

$$= e_1 \wedge e_2 \wedge e_3$$

$$T = e_{123} \tag{8.93}$$

and

$$\alpha = 180°$$

$$\hat{B} = e_{31}$$

$$T' = (\cos 90° - e_{31})e_{123}(\cos 90° + e_{31})$$

$$= -e_{31}e_{123}e_{31}$$

$$= e_{3112331}$$

$$T' = e_{123} \tag{8.94}$$

thus the trivector's sign remains unchanged.

8.11 Rotating scalars

Scalars are a grade 0 element and have no physical extent and therefore are invariant under rotations.

To summarize, we have the following rules for rotations:

$$
\begin{array}{ll}
scalars & \text{invariant} \\
vectors & v' = \hat{n}\hat{m}v\hat{m}\hat{n} \\
bivectors & B' = \hat{n}\hat{m}B\hat{m}\hat{n} \\
trivectors & T' = \hat{n}\hat{m}T\hat{m}\hat{n}.
\end{array} \tag{8.95}
$$

8.12 Rotors in exponential form

Leonhard Euler [1707–1783] discovered the following relationship between e, i, cos and sin:

$$e^{i\alpha} = \cos\alpha + i\sin\alpha \tag{8.96}$$

which takes on a rather beautiful quality when $\alpha = \pi$:

$$e^{i\pi} = -1. \tag{8.97}$$

However, as

$$R = \cos\alpha - \hat{B}\sin\alpha \qquad (8.98)$$

where $\hat{B}^2 = -1$, we can write

$$R = e^{-\hat{B}\alpha}$$

$$R = \exp(-\hat{B}\alpha). \qquad (8.99)$$

Recalling that the double reflection algorithm doubles the angle of rotation, we must compensate for this by halving the original angle:

$$R = \exp(-\hat{B}\alpha/2). \qquad (8.100)$$

Similarly,

$$\tilde{R} = \exp(\hat{B}\alpha/2). \qquad (8.101)$$

Therefore, to rotate a vector, bivector or trivector we can also use

$$M' = e^{-\hat{B}\alpha/2}Me^{\hat{B}\alpha/2}. \qquad (8.102)$$

8.13 A rotor matrix

Although geometric algebra provides new ways to calculate reflections and rotations, at the end of the day, a rotation is a rotation, and a direct relationship must exist with matrix algebra. We can demonstrate this as follows:

We begin with the bivector defining the plane $\hat{m} \wedge \hat{n}$, about which the rotation is effected, where

$$\hat{m} = m_1 e_1 + m_2 e_2 + m_3 e_3 \qquad (8.103)$$

and

$$\hat{n} = n_1 e_1 + n_2 e_2 + n_3 e_3. \qquad (8.104)$$

Therefore, we let

$$R = \hat{n}\hat{m} \qquad (8.105)$$

which can also be written as

$$R = w + x e_{23} + y e_{31} + z e_{12}. \qquad (8.106)$$

Similarly,

$$\tilde{R} = \hat{m}\hat{n} \qquad (8.107)$$

and

$$\tilde{R} = w - x e_{23} - y e_{31} - z e_{12}. \qquad (8.108)$$

where

$$w^2 + x^2 + y^2 + z^2 = 1. \qquad (8.109)$$

Therefore, given an arbitrary vector

$$v = v_1 e_1 + v_2 e_2 + v_3 e_3 \qquad (8.110)$$

we must compute the individual components $Rv_1e_1\tilde{R}$, $Rv_2e_2\tilde{R}$ and $Rv_3e_3\tilde{R}$:

$$Rv_1e_1\tilde{R} = (w + xe_{23} + ye_{31} + ze_{12})v_1e_1(w - xe_{23} - ye_{31} - ze_{12})$$
$$= v_1(we_1 + xe_{123} + ye_3 - ze_2)(w - xe_{23} - ye_{31} - ze_{12})$$
$$= v_1\left[(w^2 + x^2 - y^2 - z^2)e_1 + 2(-wz + xy)e_2 + 2(xz + wy)e_3\right] \quad (8.111)$$

but

$$w^2 + x^2 = 1 - y^2 - z^2 \quad (8.112)$$

therefore,

$$v_1' = v_1\left[\left(1 - 2(y^2 + z^2)\right)e_1 + 2(xy - wz)e_2 + 2(xz + wy)e_3\right]. \quad (8.113)$$

Next,

$$Rv_2e_2\tilde{R} = (w + xe_{23} + ye_{31} + ze_{12})v_2e_2(w - xe_{23} - ye_{31} - ze_{12})$$
$$= v_2(we_2 - xe_3 + ye_{123} - ze_1)(w - xe_{23} - ye_{31} - ze_{12})$$
$$= v_2\left[\left(2(xy + wz)e_1 + (w^2 - x^2 + y^2 - z^2)e_2 + 2(yz - wx)e_3\right)\right]. \quad (8.114)$$

Substituting

$$w^2 + y^2 = 1 - x^2 - z^2$$
$$v_2' = v_2\left[2(xy + wz)e_1 + \left(1 - 2(x^2 + z^2)\right)e_2 + 2(yz - wx)e_3\right]. \quad (8.115)$$

Next,

$$Rv_3e_3\tilde{R} = (w + xe_{23} + ye_{31} + ze_{12})v_3e_3(w - xe_{23} - ye_{31} - ze_{12})$$
$$= (we_3 + xe_2 - ye_1 + ze_{123})(w - xe_{23} - ye_{31} - ze_{12})$$
$$= v_3\left[2(xz - wy)e_1 + 2(yz + wx)e_2 + (w^2 - x^2 - y^2 + z^2)e_3\right]. \quad (8.116)$$

Substituting

$$w^2 + z^2 = 1 - x^2 - y^2$$
$$v_3' = v_3\left[2(xz + wy)e_1 + 2(yz + wx)e_2 + \left(1 - 2(x^2 + y^2)\right)e_3\right]. \quad (8.117)$$

Therefore, the rotated vector is

$$v' = Rv\tilde{R} \quad (8.118)$$

or as a matrix

$$v' = \begin{bmatrix} 1 - 2(y^2 + z^2) & 2(xy - wz) & 2(xz + wy) \\ 2(xy + wz) & 1 - 2(x^2 + z^2) & 2(yz - wx) \\ 2(xz - wy) & 2(yz + wx) & 1 - 2(x^2 + y^2) \end{bmatrix} \begin{bmatrix} v_1 \\ v_2 \\ v_3 \end{bmatrix} \quad (8.119)$$

which is the well known matrix for a quaternion!

Let's test (8.119) using the example shown in Figure 8.20, where vector v is rotated 120° about $B = n \wedge m$, where

$$n = e_1 - e_3$$

$$m = e_2 - e_3$$

$$\hat{n} = \frac{1}{\sqrt{2}}(e_1 - e_3)$$

$$\hat{m} = \frac{1}{\sqrt{2}}(e_2 - e_3). \tag{8.120}$$

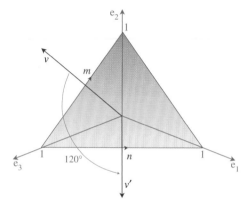

FIGURE 8.20.

Therefore,

$$R = \hat{n}\hat{m}$$

$$= \frac{1}{\sqrt{2}}(e_1 - e_3)\frac{1}{\sqrt{2}}(e_2 - e_3)$$

$$= \frac{1}{2}(e_{12} + e_{31} + e_{23} + 1)$$

$$R = \frac{1}{2} + \frac{1}{2}e_{12} + \frac{1}{2}e_{23} + \frac{1}{2}e_{31} \tag{8.121}$$

and

$$w = x = y = z = \frac{1}{2}. \tag{8.122}$$

Substituting w, x, y and z in (8.119) we have

$$v' = \begin{bmatrix} 1 - 2(y^2 + z^2) & 2(xy - wz) & 2(xz + wy) \\ 2(xy + wz) & 1 - 2(x^2 + z^2) & 2(yz - wx) \\ 2(xz - wy) & 2(yz + wx) & 1 - 2(x^2 + y^2) \end{bmatrix} \begin{bmatrix} v_1 \\ v_2 \\ v_3 \end{bmatrix}$$

$$\begin{bmatrix} 1 \\ 0 \\ 1 \end{bmatrix} = \begin{bmatrix} 0 & 0 & 1 \\ 1 & 0 & 0 \\ 0 & 1 & 0 \end{bmatrix} \begin{bmatrix} 0 \\ 1 \\ 1 \end{bmatrix} \tag{8.123}$$

and

$$v' = e_1 + e_3 \tag{8.124}$$

which is correct.

8.14 Building rotors

So far we have seen how two vectors can give rise to a rotation using

$$a' = Ra\tilde{R} \tag{8.125}$$

where

$$R = \hat{n}\hat{m} \tag{8.126}$$

and

$$\tilde{R} = \hat{m}\hat{n}. \tag{8.127}$$

Next, given a' and a, we discover how to find R and $\tilde{R}$. And instead of using a' and a, we use $\hat{b}$ and $\hat{a}$ for clarity. Note that the vectors are normalized.

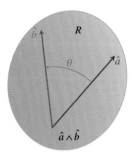

Figure 8.21.

Figure 8.21 shows vectors $\hat{a}$ and $\hat{b}$ lying in a plane defined by $\hat{a} \wedge \hat{b}$. We start by constructing a vector $\hat{n}$ mid-way between $\hat{a}$ and $\hat{b}$:

$$\hat{n} = \frac{\hat{a} + \hat{b}}{|\hat{a} + \hat{b}|} \tag{8.128}$$

and create a line l_n perpendicular to $\hat{n}$ as shown in Figure 8.22.

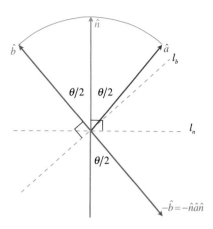

FIGURE 8.22.

Using (8.28) we create a reflection of $\hat{a}$ in l_n to create $-\hat{b}$:

$$-\hat{b} = -\hat{n}\hat{a}\hat{n}. \tag{8.129}$$

Next, we construct a line l_b perpendicular to $\hat{b}$ as shown in Figure 8.22. Again, using (8.28) we create a reflection of $-\hat{b}$ in l_b to create $\hat{b}$:

$$\hat{b} = -\hat{b}(-\hat{b})\hat{b}. \tag{8.130}$$

Substituting $-\hat{n}\hat{a}\hat{n}$ for $(-\hat{b})$ in (8.130) we have

$$\hat{b} = -\hat{b}(-\hat{n}\hat{a}\hat{n})\hat{b}$$
$$\hat{b} = \hat{b}\hat{n}\hat{a}\hat{n}\hat{b}. \tag{8.131}$$

Now we can define the rotors R and $\tilde{R}$:

$$R = \hat{b}\hat{n} \tag{8.132}$$

and

$$\tilde{R} = \hat{n}\hat{b} \tag{8.133}$$

where

$$\hat{b} = R\hat{a}\tilde{R}. \tag{8.134}$$

Next we eliminate $\hat{n}$ from the the the rotors by substituting (8.128):

$$R = \hat{b}\hat{n}$$
$$= \hat{b}\frac{\hat{a} + \hat{b}}{|\hat{a} + \hat{b}|}$$
$$R = \frac{1 + \hat{b}\hat{a}}{|\hat{a} + \hat{b}|}. \tag{8.135}$$

Similarly,

$$\tilde{R} = \hat{n}\hat{b}$$

$$= \frac{(\hat{a} + \hat{b})}{|\hat{a} + \hat{b}|}\hat{b}$$

$$\tilde{R} = \frac{1 + \hat{a}\hat{b}}{|\hat{a} + \hat{b}|}. \tag{8.136}$$

The denominator in (8.135) and (8.136) can be resolved into an alternate form using the following subterfuge:

Figure 8.23 shows the geometry associated with the plane $\hat{a} \wedge \hat{b}$. From the figure we see that

$$\frac{\lambda}{|\hat{a}|} = \cos(\theta/2)$$

$$\lambda = \cos(\theta/2). \tag{8.137}$$

Using the half-angle identity

$$\lambda = \cos(\theta/2) = \sqrt{\frac{1 + \cos\theta}{2}}. \tag{8.138}$$

But

$$|\hat{a} + \hat{b}| = 2\lambda$$

$$= 2\sqrt{\frac{1 + \cos\theta}{2}}$$

$$|\hat{a} + \hat{b}| = \sqrt{2(1 + \cos\theta)}. \tag{8.139}$$

And as

$$\hat{a} \cdot \hat{b} = \cos\theta$$

$$|\hat{a} + \hat{b}| = \sqrt{2(1 + \hat{a} \cdot \hat{b})} \tag{8.140}$$

therefore,

$$R = \frac{1 + \hat{b}\hat{a}}{\sqrt{2(1 + \hat{a} \cdot \hat{b})}}. \tag{8.141}$$

Similarly,

$$\tilde{R} = \frac{1 + \hat{a}\hat{b}}{\sqrt{2(1 + \hat{a} \cdot \hat{b})}}. \tag{8.142}$$

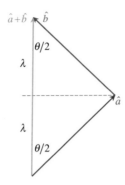

FIGURE 8.23.

Let's see how this rotor behaves with various vectors. For example, Figure 8.24 shows a scenario where the angle of rotation is 90° and the vectors are

$$\hat{a} = e_1$$
$$\hat{b} = e_2 \tag{8.143}$$

and

$$R = \frac{1 + e_2 e_1}{\sqrt{2(1 + e_1 \cdot e_2)}}$$
$$R = \frac{1 + e_{21}}{\sqrt{2}} \tag{8.144}$$

and

$$\tilde{R} = \frac{1 + e_1 e_2}{\sqrt{2(1 + e_1 \cdot e_2)}}$$
$$\tilde{R} = \frac{1 + e_{12}}{\sqrt{2}}. \tag{8.145}$$

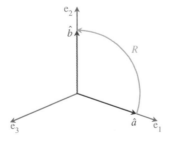

FIGURE 8.24.

We begin by rotating $\hat{a} = e_1$:

$$R\hat{a}\tilde{R} = \left(\frac{1 + e_{21}}{\sqrt{2}}\right) e_1 \left(\frac{1 + e_{12}}{\sqrt{2}}\right)$$

$$= \frac{1}{2}(e_1 + e_2)(1 + e_{12})$$

$$= \frac{1}{2}(e_1 + e_2 + e_2 - e_1)$$

$$= e_2$$

$$R\hat{a}\tilde{R} = \hat{b}. \tag{8.146}$$

As expected, $\hat{a}$ is rotated to $\hat{b}$.

Now let's chose a vector half-way between $\hat{a}$ and $\hat{b}$: $e_1 + e_2$

$$R(e_1 + e_2)\tilde{R} = \left(\frac{1 + e_{21}}{\sqrt{2}}\right) (e_1 + e_2) \left(\frac{1 + e_{12}}{\sqrt{2}}\right)$$

$$= \frac{1}{2}(e_1 + e_2 + e_2 - e_1)(1 + e_{12})$$

$$= \frac{1}{2}(2e_2 - 2e_1)$$

$$R(e_1 + e_2)\tilde{R} = -e_1 + e_2. \tag{8.147}$$

Thus the vector has been rotated 90°.

Next, we rotate $\hat{b} = e_2$:

$$R\hat{b}\tilde{R} = \left(\frac{1 + e_{21}}{\sqrt{2}}\right) e_2 \left(\frac{1 + e_{12}}{\sqrt{2}}\right)$$

$$= \frac{1}{2}(e_2 - e_1)(1 + e_{12})$$

$$= \frac{1}{2}(e_2 - e_1 - e_1 - e_2)$$

$$R\hat{b}\tilde{R} = -e_1. \tag{8.148}$$

Thus $\hat{b}$ has been rotated to 90° to $-e_1$.

Next, we rotate $-e_1$:

$$R(-e_1)\tilde{R} = \left(\frac{1 + e_{21}}{\sqrt{2}}\right) (-e_1) \left(\frac{1 + e_{12}}{\sqrt{2}}\right)$$

$$= \frac{1}{2}(-e_1 - e_2)(1 + e_{12})$$

$$= \frac{1}{2}(-e_1 - e_2 - e_2 + e_1)$$

$$R(-e_1)\tilde{R} = -e_2. \tag{8.149}$$

Thus, this, too, is rotated 90°.

Before moving on, let's consider the following algebra:
Using (8.135) we have

$$R\hat{a} = \left(\frac{1 + \hat{b}\hat{a}}{\sqrt{2(1 + \hat{a} \cdot \hat{b})}} \right) \hat{a}$$

$$R\hat{a} = \frac{\hat{a} + \hat{b}}{|\hat{a} + \hat{b}|}. \tag{8.150}$$

Similarly,

$$\hat{a}\tilde{R} = \hat{a} \left(\frac{1 + \hat{a}\hat{b}}{|\hat{a} + \hat{b}|} \right)$$

$$\hat{a}\tilde{R} = \frac{\hat{a} + \hat{b}}{|\hat{a} + \hat{b}|} \tag{8.151}$$

therefore,

$$R\hat{a} = \hat{a}\tilde{R} \tag{8.152}$$

which reveals that pre-multiplying a vector by a rotor is equivalent to post-multiplying it by the inverse rotor, which leads to

$$R^2\hat{a} = R\hat{a}\tilde{R} = \hat{b}. \tag{8.153}$$

Which means that having found R^2, a vector can be rotated through a similar angle in the same plane using (8.153). So let's find R^2 for the above rotor:

$$R^2\hat{a} = \hat{b} \tag{8.154}$$

therefore,

$$R^2\hat{a}\hat{a}^{-1} = \hat{b}\hat{a}^{-1}$$

$$R^2 = \hat{b}\hat{a}. \tag{8.155}$$

(Remember, the inverse of a unit vector is the original vector)
For the previous example

$$R^2 = e_2 e_1^{-1}$$

$$= e_2 e_1$$

$$R^2 = -e_{12}. \tag{8.156}$$

Any vector can now be rotated 90° within the plane containing $\hat{a}$ and $\hat{b}$ using R^2. For example, if

$$c = 3e_1 + 2e_2$$

then

$$c' = R^2 c$$
$$= -e_{12}(3e_1 + 2e_2)$$
$$c' = 3e_2 - 2e_1 \tag{8.157}$$

which is correct.

To conclude this section on building rotors, let's design a rotor to rotate $135°$ within the plane e_{31}. Figure 8.25 shows two vectors

$$\hat{a} = e_3 \tag{8.158}$$

and

$$\hat{b} = \frac{1}{\sqrt{2}}(e_1 - e_3) \tag{8.159}$$

which provide the plane and angle of rotation.

If we now rotate the vector

$$c = -e_1 \tag{8.160}$$

we have

$$c' = \hat{b}\hat{a}c$$
$$= \frac{1}{\sqrt{2}}(e_1 - e_3)e_3(-e_1)$$
$$= \frac{1}{\sqrt{2}}(e_{13} - 1)(-e_1)$$
$$c' = \frac{1}{\sqrt{2}}(e_1 + e_3) \tag{8.161}$$

which is also correct.

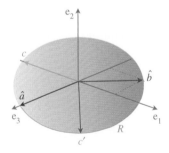

FIGURE 8.25.

8.15 Interpolating rotors

Interpolating scalars is a trivial exercise and is readily implemented using the linear interpolant

$$\alpha = \alpha_1(1 - \lambda) + \alpha_2\lambda \qquad [\alpha, \lambda \in \mathbb{R}] \tag{8.162}$$

where

$$0 \leq \lambda \leq 1 \tag{8.163}$$

and there is no reason why it cannot be used for vectors, apart from the fact that the interpolated vector could collapse to a null vector.

To preserve a vector's geometric integrity we must manage both its orientation and magnitude. To achieve this, consider the scenario shown in Figure 8.26 where two unit vectors $\hat{v}_1$ and $\hat{v}_2$ are separated by an angle θ.

The interpolated vector $\hat{v}$ can be defined as a linear proportion of $\hat{v}_1$ and $\hat{v}_2$:

$$\hat{v} = \alpha\hat{v}_1 + \beta\hat{v}_2. \tag{8.164}$$

Let's define α and β such that they are a function of the angle θ. Vector $\hat{v}$ is $\lambda\theta$ from $\hat{v}_1$ and $(1-\lambda)\theta$ from $\hat{v}_2$, and it is evident from Figure 8.26 that using the sine rule:

$$\frac{\alpha}{\sin((1 - \lambda)\theta)} = \frac{\beta}{\sin(\lambda\theta)} \tag{8.165}$$

and furthermore,

$$\varepsilon_1 + \varepsilon_2 = 1 \tag{8.166}$$

where

$$\varepsilon_1 = \alpha \cos(\lambda\theta)$$
$$\varepsilon_2 = \beta \cos((1 - \lambda)\theta). \tag{8.167}$$

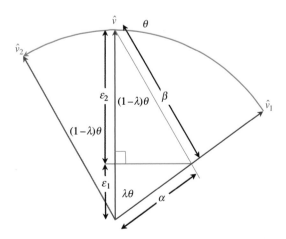

FIGURE 8.26.

From (8.165)

$$\beta = \frac{\alpha \sin(\lambda\theta)}{\sin((1-\lambda)\theta)} \qquad (8.168)$$

and from (8.166) we obtain

$$\alpha \cos(\lambda\theta) + \beta \cos((1-\lambda)\theta) = 1 \qquad (8.169)$$

substituting (8.168) we have

$$\alpha \cos(\lambda\theta) + \frac{\alpha \sin(\lambda\theta)}{\sin((1-\lambda)\theta)} \cos((1-\lambda)\theta) = 1. \qquad (8.170)$$

Solving for α and β we obtain:

$$\alpha = \frac{\sin((1-\lambda)\theta)}{\sin\theta}$$

$$\beta = \frac{\sin(\lambda\theta)}{\sin\theta} \qquad (8.171)$$

therefore,

$$\hat{v} = \frac{\sin((1-\lambda)\theta)}{\sin\theta}\hat{v}_1 + \frac{\sin(\lambda\theta)}{\sin\theta}\hat{v}_2. \qquad (8.172)$$

Such an interpolant is called a *slerp* (spherical-linear interpolant).

Although (8.172) has been designed to work with unit vectors, it behaves correctly with vectors of arbitrary magnitude. Furthermore, this slerp can also be used to interpolate quaternions and rotors using:

$$\hat{R} = \frac{\sin((1-\lambda)\theta/2)}{\sin(\theta/2)}\hat{R}_1 + \frac{\sin(\lambda\theta/2)}{\sin(\theta/2)}\hat{R}_2 \qquad [0 \le \lambda \le 1] \qquad (8.173)$$

where θ is the angle between the rotors.

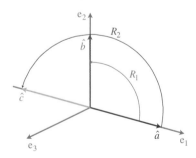

Figure 8.27.

To keep the following algebra simple, we will construct two easy rotors. For example, Figure 8.27 shows three unit vectors $\hat{a}$, $\hat{b}$ and $\hat{c}$.

Rotor R_1 rotates $\hat{a}$ to $\hat{b}$, whilst rotor R_2 rotates $\hat{a}$ to $\hat{c}$. The angular difference between the rotors is $\theta = 90°$.

Therefore, if

$$\hat{a} = e_1$$
$$\hat{b} = e_2$$
$$\hat{c} = -e_1$$
$$\theta = 90° \tag{8.174}$$

then

$$R_1 = \hat{b}\hat{a}$$
$$= \cos 45° - e_{12} \sin 45°$$
$$R_1 = \frac{\sqrt{2}}{2} - e_{12}\frac{\sqrt{2}}{2}$$
$$R_2 = \hat{c}\hat{a}$$
$$= \cos 90° - e_{12} \sin 90°$$
$$R_2 = -e_{12}. \tag{8.175}$$

Using (8.173)

$$\hat{R} = \frac{\sin((1-\lambda)90°/2)}{\sin(90°/2)}\hat{R}_1 + \frac{\sin(\lambda 90°/2)}{\sin(90°/2)}\hat{R}_2$$
$$\hat{R} = \frac{\sin((1-\lambda)45°)}{\sin(45°)}\hat{R}_1 + \frac{\sin(\lambda 45°)}{\sin(45°)}\hat{R}_2 \tag{8.176}$$

thus when $\lambda = 0, \hat{R} = \hat{R}_1$, and when $\lambda = 1, \hat{R} = \hat{R}_2$.

Let's compute $\hat{R}$ when $\lambda = 0.5$:

$$\hat{R} = \frac{\sin 22.5°}{\sin 45°}\left(\frac{\sqrt{2}}{2} - e_{12}\frac{\sqrt{2}}{2}\right) + \frac{\sin 22.5°}{\sin 45°}(-e_{12}).$$

$$= \delta\left(\frac{\sqrt{2}}{2} - e_{12}\frac{\sqrt{2}}{2}\right) + \delta(-e_{12}) \tag{8.177}$$

where

$$\delta \simeq 0.5411961.$$

Therefore,

$$\hat{R} \simeq 0.3826 - 0.9237e_{12} \tag{8.178}$$

and

$$\tilde{\hat{R}} \simeq 0.3826 + 0.9237e_{12}. \tag{8.179}$$

Consequently,

$$R = \hat{R}a\tilde{\hat{R}}$$

$$\simeq (0.3826 - 0.9237e_{12})e_1(0.3826 + 0.9237e_{12})$$

$$R \simeq -0.707e_1 + 0.7071e_2 \qquad\qquad (8.180)$$

which is half-way between the two rotors.

8.16 Summary

In this chapter we have seen how geometric algebra unifies reflections and rotations. Furthermore, it is possible to understand how excited Hamilton must have been when he stumbled across the rules for quaternion products, even if they were not the complete answer. He knew that 2D complex numbers could be rotated in the complex plane using a single rotor and innocently believed that a similar mechanism existed in 3D. Eventually, he realised that a 3D vector had to be sandwiched between two rotors to effect a rotation. Now, it is understood that quaternions are just a feature of geometric algebra.

But what of Grassmann? He had been trying to tell the world that his algebra of extensions was the answer, and although his voice was heard, his message was not understood. This in itself should be a lesson to all mathematicians: keep it simple!

9 Applied Geometric Algebra

9.1 Introduction

In this chapter we explore how geometric algebra can be used to resolve a variety of 2D and 3D geometric problems. In particular, we discover that the vector-bivector inner product is useful for rotating vectors and that the vector outer product is very useful in resolving problems where oriented areas are involved. To begin, we take a look at some simple geometric proofs and start with the familiar sine rule.

9.2 Sine rule

The sine rule states that for any triangle $\triangle ABC$ with angles α, β and χ, and respective opposite sides a, b and c then

$$\frac{a}{\sin \alpha} = \frac{b}{\sin \beta} = \frac{c}{\sin \chi}.$$
(9.1)

This rule can be proved using the outer product of two vectors, which we know incorporates the sine of the angle between two vectors:

$$|a \wedge b| = |a||b| \sin \alpha.$$
(9.2)

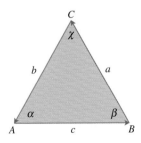

FIGURE 9.1.

J. Vince, *Geometric Algebra: An Algebraic System for Computer Games and Animation,*
© Springer-Verlag London Limited 2009

With reference to Figure 9.1, the triangle's sides are defined by vectors a, b and c, which permits us to state the triangle's area as

$$\text{area of } \triangle ABC = \frac{1}{2}|-c \wedge a| = \frac{1}{2}|c||a| \sin \beta$$

$$\text{area of } \triangle BCA = \frac{1}{2}|-a \wedge b| = \frac{1}{2}|a||b| \sin \chi$$

$$\text{area of } \triangle CAB = \frac{1}{2}|-b \wedge c| = \frac{1}{2}|b||c| \sin \alpha \tag{9.3}$$

which means that

$$|c||a| \sin \beta = |a||b| \sin \chi = |b||c| \sin \alpha \tag{9.4}$$

and

$$\frac{|a|}{\sin \alpha} = \frac{|b|}{\sin \beta} = \frac{|c|}{\sin \chi} \tag{9.5}$$

which confirms (9.151).

9.3 Cosine rule

The cosine rule states that for any triangle $\triangle ABC$ with sides a, b and c then

$$a^2 = b^2 + c^2 - 2bc \cos \alpha \tag{9.6}$$

where α is the angle between b and c.

Although this is an easy rule to prove using simple trigonometry, the geometric algebra solution is even easier.

Figure 9.2 shows a triangle $\triangle ABC$ constructed from vectors a, b and c, such that b and c have orientations to form an inner or outer product.

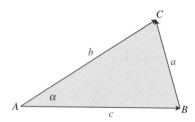

FIGURE 9.2.

From Figure 9.2

$$a = b - c \tag{9.7}$$

squaring (9.7)

$$a^2 = b^2 + c^2 - (bc + cb). \tag{9.8}$$

But

$$bc + cb = 2b \cdot c = 2|b||c| \cos \alpha \qquad (9.9)$$

and using the modulus axiom (3.28)

$$|a|^2 = |b|^2 + |c|^2 - 2|b||c| \cos \alpha. \qquad (9.10)$$

9.4 Inscribed circle of a triangle

The inscribed circle of a triangle is positioned such that it touches every side of the triangle, and we can find this position as follows. Figure 9.3 shows a triangle $\triangle ABC$ constructed from vectors a, b and c. The centre P is identified by the common intersection of the angles' bisectors.

The problem is solved in 3D to obtain a general solution.

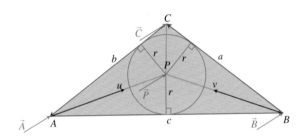

FIGURE 9.3.

We define the following unit vectors:

$$\hat{a} = \frac{a}{|a|} \quad \hat{b} = \frac{b}{|b|} \quad \hat{c} = \frac{c}{|c|} \qquad (9.11)$$

and for convenience make the bisectors

$$u = \hat{c} - \hat{b} \quad v = \hat{a} - \hat{c}. \qquad (9.12)$$

Therefore,

$$\vec{P} = \vec{A} + \lambda u$$
$$\vec{P} = \vec{B} + \varepsilon v \qquad (9.13)$$

and

$$\lambda u - \varepsilon v = \vec{B} - \vec{A}$$
$$\lambda u - \varepsilon v = c. \qquad (9.14)$$

Taking the outer product of (9.14) using v we get

$$\lambda(u \wedge v) - \varepsilon(v \wedge v) = c \wedge v \qquad (9.15)$$

but because $v \wedge v = 0$

$$
\begin{aligned}
\lambda &= \frac{c \wedge v}{u \wedge v} \\
&= \frac{c \wedge (\hat{a} - \hat{c})}{(\hat{c} - \hat{b}) \wedge (\hat{a} - \hat{c})} \\
&= \frac{c \wedge \hat{a}}{(\hat{c} \wedge \hat{a}) + (\hat{a} \wedge \hat{b}) + (\hat{b} \wedge \hat{c})} \\
&= \frac{c \wedge \frac{a}{|a|}}{\left(\frac{c}{|c|} \wedge \frac{a}{|a|}\right) + \left(\frac{a}{|a|} \wedge \frac{b}{|b|}\right) + \left(\frac{b}{|b|} \wedge \frac{c}{|c|}\right)} \\
&= \frac{|b||c|(c \wedge a)}{|b|(c \wedge a) + |c|(a \wedge b) + |a|(b \wedge c)}
\end{aligned}
\tag{9.16}
$$

but because

$$
c \wedge a = a \wedge b = b \wedge c
$$

$$
\lambda = \frac{|b||c|}{|a| + |b| + |c|}.
\tag{9.17}
$$

Therefore,

$$
\begin{aligned}
\vec{P} &= \vec{A} + \lambda u \\
&= \vec{A} + \left(\frac{|b||c|}{|a| + |b| + |c|}\right)\left(\frac{c}{|c|} - \frac{b}{|b|}\right) \\
&= \vec{A} + \left(\frac{|b||c|}{|a| + |b| + |c|}\right)\left(\frac{\vec{B} - \vec{A}}{|c|} - \frac{\vec{A} - \vec{C}}{|b|}\right) \\
\vec{P} &= \frac{|a|\vec{A} + |b|\vec{B} + |c|\vec{C}}{|a| + |b| + |c|}.
\end{aligned}
\tag{9.18}
$$

To find the radius r we compute the area of the triangle:

$$
\begin{aligned}
\text{area } \triangle ABC &= \frac{1}{2}r|a| + \frac{1}{2}r|b| + \frac{1}{2}r|c| \\
&= \frac{1}{2}r(|a| + |b| + |c|).
\end{aligned}
\tag{9.19}
$$

But the area can also be defined in terms of the outer product using any two sides:

$$
\text{area } \triangle ABC = \frac{1}{2}|a \wedge b|
\tag{9.20}
$$

therefore,

$$
\frac{1}{2}|a \wedge b| = \frac{1}{2}r(|a| + |b| + |c|)
\tag{9.21}
$$

and

$$r = \frac{|a \wedge b|}{|a| + |b| + |c|}. \tag{9.22}$$

For example, given

$$\vec{A} = \begin{bmatrix} 0 \\ 0 \end{bmatrix} \quad \vec{B} = \begin{bmatrix} 2 \\ 0 \end{bmatrix} \quad \vec{C} = \begin{bmatrix} 0 \\ 2 \end{bmatrix} \tag{9.23}$$

with

$$a = \begin{bmatrix} -2 \\ 2 \end{bmatrix} \quad b = \begin{bmatrix} 0 \\ 2 \end{bmatrix} \quad c = \begin{bmatrix} 2 \\ 0 \end{bmatrix} \tag{9.24}$$

then

$$|a| = \sqrt{8} \quad |b| = 2 \quad |c| = 2 \tag{9.25}$$

and

$$\vec{P} = \frac{\sqrt{8} \begin{bmatrix} 0 \\ 0 \end{bmatrix} + 2 \begin{bmatrix} 2 \\ 0 \end{bmatrix} + 2 \begin{bmatrix} 0 \\ 2 \end{bmatrix}}{\sqrt{8} + 2 + 2}$$

$$\vec{P} = \begin{bmatrix} \frac{4}{4+\sqrt{8}} \\ \frac{4}{4+\sqrt{8}} \end{bmatrix} \tag{9.26}$$

and

$$r = \frac{\left\| \begin{matrix} -2 & 2 \\ 0 & 2 \end{matrix} \right\|}{\sqrt{8} + 2 + 2}$$

$$r = \frac{4}{4 + \sqrt{8}}. \tag{9.27}$$

9.5 Circumscribed circle of a triangle

Next, we derive a solution for the circumscribed circle of a 3D triangle. Given any triangle, the circumscribed circle is positioned such that it touches every vertex. We can find the circle's radius r and centre P as follows:

Figure 9.4 shows a triangle $\triangle ABC$ with position vectors $\vec{A}$, $\vec{B}$ and $\vec{C}$, and sides defined by vectors a, b and c.

Point P is positioned such that it is distance r (the circumscribed radius) from A, B and C. Consequently, $\triangle ABP$, $\triangle BCP$ and $\triangle CAP$ are isosceles triangles and perpendiculars from P bisect the sides.

Let h be the perpendicular distance B of from the line CA.

From the chord theorem

$$\angle APB = 2\angle ACB = 2\alpha \tag{9.28}$$

and

$$\sin \alpha = \frac{|c|}{2r} = \frac{h}{|a|}. \tag{9.29}$$

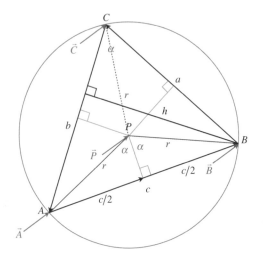

Figure 9.4.

Therefore,

$$h = \frac{|a||c|}{2r}. \tag{9.30}$$

The area of $\triangle ABC$ equals

$$\begin{aligned}
\text{area of } \triangle ABC &= \frac{1}{2}|b|h \\
&= \frac{|b||a||c|}{4r} \\
&= \frac{|a||b||c|}{4r}. \tag{9.31}
\end{aligned}$$

But the area of $\triangle ABC$ also equals

$$\text{area of } \triangle ABC = \frac{1}{2}|a \wedge b| \tag{9.32}$$

therefore,

$$\frac{|a||b||c|}{4r} = \frac{1}{2}|a \wedge b| \tag{9.33}$$

and the radius of the circumscribed circle is

$$r = \frac{|a||b||c|}{2|a \wedge b|}. \tag{9.34}$$

This corresponds with the well-known formula

$$r = \frac{abc}{4 \times \text{area} \triangle ABC}. \tag{9.35}$$

The position of P is identified by analyzing triangle $\triangle ABP$ as shown in Figure 9.5.

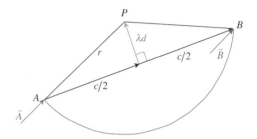

FIGURE 9.5.

Let d be perpendicular vector from AB towards P where $|d| = |c|$ therefore,

$$r^2 = \left|\frac{c}{2}\right|^2 + |\lambda d|^2$$

$$= \frac{|c|^2}{4} + \lambda^2 |c|^2 \tag{9.36}$$

and

$$\lambda = \sqrt{\frac{r^2}{|c|^2} - \frac{1}{4}} \tag{9.37}$$

and

$$\vec{P} = \vec{A} + \frac{c}{2} + \lambda d. \tag{9.38}$$

We now need to compute d which is effected by using the bivector

$$T = a \wedge (-c) \tag{9.39}$$

which rotates c through $90°$ within its plane using

$$d = \frac{c \cdot T}{|T|}. \tag{9.40}$$

Note that the bivector T is normalized to prevent any scaling.

Finally,

$$\vec{P} = \vec{A} + \frac{c}{2} + \frac{\lambda c \cdot T}{|T|}. \tag{9.41}$$

Let's test these formulae with an example.

Figure 9.6 shows a triangle intersecting the three axes at $(1, 0, 0)$, $(0, 1, 0)$, $(0, 0, 1)$.

The triangle is formed from vectors

$$a = -e_1 + e_2$$

$$b = -e_2 + e_3$$

$$c = e_1 - e_3 \tag{9.42}$$

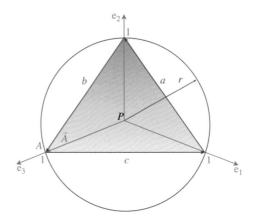

FIGURE 9.6.

where

$$|a| = |b| = |c| = \sqrt{2}. \tag{9.43}$$

Using (9.34)

$$r = \frac{|a||b||c|}{2|a \wedge b|}$$

$$= \frac{\sqrt{2}\sqrt{2}\sqrt{2}}{2|(-e_1 + e_2) \wedge (-e_2 + e_3)|}$$

$$= \frac{\sqrt{2}}{|e_1 \wedge e_2 - e_1 \wedge e_3 + e_2 \wedge e_3|}$$

$$r = \frac{\sqrt{2}}{\sqrt{3}} = \sqrt{\frac{2}{3}}. \tag{9.44}$$

Next we calculate λ using (9.37)

$$\lambda = \sqrt{\frac{r^2}{|c|^2} - \frac{1}{4}}$$

$$= \sqrt{\frac{2/3}{2} - \frac{1}{4}}$$

$$\lambda = \sqrt{\frac{1}{12}}. \tag{9.45}$$

Next we calculate the bivector T using (9.39)

$$T = a \wedge (-c)$$

$$= (-e_1 + e_2) \wedge (-e_1 + e_3)$$

$$T = e_{31} + e_{12} + e_{23}. \tag{9.46}$$

Next we calculate $\vec{P}$ using (9.41)

$$\vec{P} = \vec{A} + \frac{c}{2} + \frac{\lambda c \cdot T}{|T|}$$

$$= e_3 + \frac{e_1 - e_3}{2} + \frac{(e_1 - e_3) \cdot (e_{12} + e_{23} + e_{31})}{\sqrt{12}\sqrt{3}}$$

$$= e_3 + \frac{e_1}{2} - \frac{e_3}{2} + \frac{1}{6}(e_2 - e_3 + e_2 - e_1)$$

$$= e_3 + \frac{e_1}{2} - \frac{e_3}{2} + \frac{1}{6}(-e_1 + 2e_2 - e_3)$$

$$= \left(\frac{1}{2} - \frac{1}{6}\right) e_1 + \left(\frac{1}{3}\right) e_2 + \left(\frac{1}{2} - \frac{1}{6}\right) e_3$$

$$\vec{P} = \frac{e_1}{3} + \frac{e_2}{3} + \frac{e_3}{3}. \tag{9.47}$$

9.6 A point perpendicular to a point on a line

Figure 9.7 shows a scenario where a line with direction vector $\hat{v}$ passes through a point T. The objective is to locate another point P perpendicular to $\hat{v}$ and a distance δ from T.

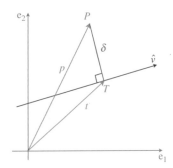

FIGURE 9.7.

The solution is found by post-multiplying $\hat{v}$ by the pseudoscalar e_{12} which rotates $\hat{v}$ through an angle of 90°. As $\hat{v}$ is a unit vector

$$\overrightarrow{TP} = \delta\hat{v}e_{12} \tag{9.48}$$

therefore,

$$p = t + \overrightarrow{TP}$$

$$p = t + \delta\hat{v}e_{12}. \tag{9.49}$$

For example, Figure 9.8 shows a 2D scenario where

$$\hat{v} = \frac{1}{\sqrt{2}}(e_1 + e_2)$$

$$T = (4, 1)$$

$$t = 4e_1 + e_2$$

$$\delta = \sqrt{32}. \tag{9.50}$$

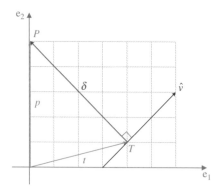

FIGURE 9.8.

Using (9.49)

$$p = t + \delta \hat{v} e_{12}$$

$$= 4e_1 + e_2 + \sqrt{32}\frac{1}{\sqrt{2}}(e_1 + e_2)e_{12}$$

$$= 4e_1 + e_2 + 4e_2 - 4e_1$$

$$p = 5e_2 \tag{9.51}$$

and

$$P = (0, 5). \tag{9.52}$$

If p is required on the opposite side of the line, we pre-multiply $\hat{v}$ by e_{12}:

$$p = t + \delta e_{12}\hat{v} \tag{9.53}$$

which is the same as reversing the sign of δ.

9.7　A point rotated an angle relative to a point on a line

Figure 9.9 shows a line with unit direction vector $\hat{v}$ that intersects a point T. The objective is to calculate a point P, distance δ from T such that $\overrightarrow{TP}$ subtends an angle α with $\hat{v}$.

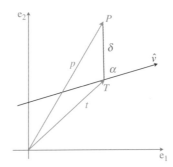

FIGURE 9.9.

We know that a vector

$$\hat{v} = v_1 e_1 + v_2 e_2 \tag{9.54}$$

is rotated α in the e_{12} plane using

$$v' = R\hat{v}\tilde{R} \tag{9.55}$$

where

$$R = \cos \alpha/2 - \sin \alpha/2 e_{12}$$
$$\tilde{R} = \cos \alpha/2 + \sin \alpha/2 e_{12} \tag{9.56}$$

which means that

$$\overrightarrow{TP} = \delta R\hat{v}\tilde{R} \tag{9.57}$$

and

$$p = t + \delta R\hat{v}\tilde{R}. \tag{9.58}$$

To demonstrate (9.58) we choose an easy example to keep the algebra simple.
Figure 9.10 shows a scenario where P is two units from T.

$$\hat{v} = e_1$$
$$T = (3, 2)$$
$$t = 3e_1 + 2e_2$$
$$\alpha = 90°$$
$$\delta = 2. \tag{9.59}$$

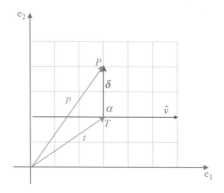

FIGURE 9.10.

Therefore,

$$R = \cos 45° - \sin 45° e_{12}$$

$$\tilde{R} = \cos 45° + \sin 45° e_{12}$$

$$\overrightarrow{TP} = 2 \left(\frac{\sqrt{2}}{2} - \frac{\sqrt{2}}{2} e_{12} \right) e_1 \left(\frac{\sqrt{2}}{2} + \frac{\sqrt{2}}{2} e_{12} \right)$$

$$= 2 \left(\frac{\sqrt{2}}{2} e_1 + \frac{\sqrt{2}}{2} e_2 \right) \left(\frac{\sqrt{2}}{2} + \frac{\sqrt{2}}{2} e_{12} \right)$$

$$= 2 \left(\frac{e_1}{2} + \frac{e_2}{2} + \frac{e_2}{2} - \frac{e_1}{2} \right)$$

$$\overrightarrow{TP} = 2e_2 \tag{9.60}$$

and

$$p = t + \overrightarrow{TP}$$

$$= 3e_1 + 2e_2 + 2e_2$$

$$p = 3e_1 + 4e_2 \tag{9.61}$$

and

$$P = (3, 4). \tag{9.62}$$

9.8 Reflecting a vector about another vector

Reflecting a vector about another vector happens to be a rather easy problem for geometric algebra. Figure 9.11 shows the scenario where we see vector a reflected about the normal to a line with direction vector $\hat{v}$.

We begin by calculating $\hat{m}$:

$$\hat{m} = \hat{v}e_{12} \tag{9.63}$$

then reflecting a about $\hat{m}$:

$$a' = \hat{m}a\hat{m}$$

$$a' = \hat{v}e_{12}a\hat{v}e_{12}. \tag{9.64}$$

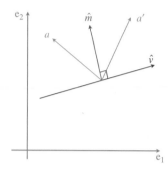

FIGURE 9.11.

As an illustration, consider the scenario shown in Figure 9.12 where

$$\hat{v} = \frac{1}{\sqrt{2}}(e_1 + e_2) \tag{9.65}$$

and

$$a = -e_1. \tag{9.66}$$

Therefore, using (9.63)

$$\hat{m} = \frac{1}{\sqrt{2}}(e_1 + e_2)e_{12}$$

$$\hat{m} = \frac{1}{\sqrt{2}}(e_2 - e_1) \tag{9.67}$$

and using (9.64)

$$a' = \frac{1}{\sqrt{2}}(e_2 - e_1)(-e_1)\frac{1}{\sqrt{2}}(e_2 - e_1)$$

$$= \frac{1}{2}(e_{12} + 1)(e_2 - e_1)$$

$$= \frac{1}{2}(e_1 + e_2 + e_2 - e_1)$$

$$a' = e_2. \tag{9.68}$$

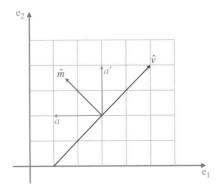

FIGURE 9.12.

9.9 The position and distance of the nearest point on a line to a point

This problem is best understood by referring to Figure 9.13 which shows a line with direction vector $\hat{v}$. P is a point in space, and the objective is to find a point Q on the line where PQ is a minimum. By simple geometry, PQ must be perpendicular to the original line.

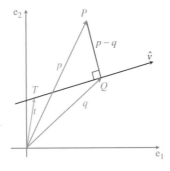

FIGURE 9.13.

T is a point on the line with position vector t.
P is a reference point with position vector P.
$\hat{v}$ is the line's direction vector.

The traditional approach is as follows:

$$q = t + \lambda\hat{v} \qquad\qquad [\lambda \in \mathbb{R}] \qquad\qquad (9.69)$$

$$\overrightarrow{QP} = p - q \qquad\qquad \text{and is } \perp \text{ to } \hat{v} \qquad\qquad (9.70)$$

therefore,

$$(p - q) \cdot \hat{v} = 0$$
$$p \cdot \hat{v} = q \cdot \hat{v}$$
$$= (t + \lambda \hat{v}) \cdot \hat{v}$$
$$p \cdot \hat{v} = t \cdot \hat{v} + \lambda \tag{9.71}$$

and

$$\lambda = (p - t) \cdot \hat{v} \tag{9.72}$$

therefore,

$$q = t + [(p - t) \cdot \hat{v}]\hat{v} \tag{9.73}$$

and

$$|\overrightarrow{QP}| = |p - q|. \tag{9.74}$$

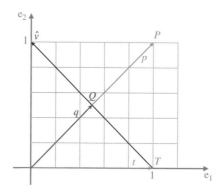

FIGURE 9.14.

As an example, Figure 9.14 shows a line represented by

$$q = t + \lambda \hat{v} \tag{9.75}$$

with

$$P = (1, 1)$$
$$p = e_1 + e_2$$
$$T = (1, 0)$$
$$t = e_1$$
$$\hat{v} = \frac{1}{\sqrt{2}}(-e_1 + e_2). \tag{9.76}$$

Therefore, using (9.73)

$$q = t + \left[(p - t) \cdot \hat{v}\right] \hat{v}$$

$$= e_1 + \left[(e_1 + e_2 - e_1) \cdot \frac{1}{\sqrt{2}}(-e_1 + e_2)\right] \frac{1}{\sqrt{2}}(-e_1 + e_2)$$

$$= e_1 + \frac{1}{2}\left[e_2 \cdot (e_2 - e_1)\right](e_2 - e_1)$$

$$= e_1 + \frac{1}{2}(e_2 - e_1)$$

$$q = \frac{e_1}{2} + \frac{e_2}{2}$$

$$Q = \left(\frac{1}{2}, \frac{1}{2}\right). \tag{9.77}$$

Using (9.74)

$$|\overrightarrow{QP}| = |p - q|$$

$$= \left|e_1 + e_2 - \frac{1}{2}e_1 - \frac{1}{2}e_2\right|$$

$$= \left|\frac{e_1}{2} + \frac{e_2}{2}\right|$$

$$|\overrightarrow{QP}| = \frac{1}{\sqrt{2}}. \tag{9.78}$$

Now let's solve the same problem using geometric algebra.

We begin by forming the geometric product $q\hat{v}$

$$q\hat{v} = q \cdot \hat{v} + q \wedge \hat{v}. \tag{9.79}$$

As $\overrightarrow{QP}$ is $\perp$ to $\hat{v}$

$$q \cdot \hat{v} = p \cdot \hat{v}. \tag{9.80}$$

The nature of the outer product is such that

$$q \wedge \hat{v} = t \wedge \hat{v} \tag{9.81}$$

therefore,

$$q\hat{v} = p \cdot \hat{v} + t \wedge \hat{v}. \tag{9.82}$$

Multiplying throughout by $\hat{v}^{-1}$ we isolate q

$$q = (p \cdot \hat{v} + t \wedge \hat{v})\hat{v}. \tag{9.83}$$

(Remember that $\hat{v}^{-1} = \hat{v}$)

With reference to Figure 9.14

$$q = \left[(e_1 + e_2) \cdot \frac{1}{\sqrt{2}} (-e_1 + e_2) + e_1 \wedge \frac{1}{\sqrt{2}} (-e_1 + e_2) \right] \frac{1}{\sqrt{2}} (-e_1 + e_2)$$

$$= \left(0 + \frac{1}{\sqrt{2}} e_{12} \right) \frac{1}{\sqrt{2}} (-e_1 + e_2)$$

$$q = \frac{e_1}{2} + \frac{e_2}{2}$$

$$Q = \left(\frac{1}{2}, \frac{1}{2} \right). \tag{9.84}$$

9.10 A line equidistant from two points

This problem only arises in 2D, where we have two points connected by a straight line. The task is to bisect the connecting line and establish a perpendicular vector, as shown in Figure 9.15.

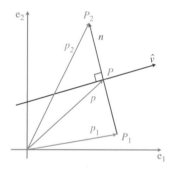

FIGURE 9.15.

From Figure 9.15

$$n = p_2 - p_1 \tag{9.85}$$

and

$$p = \frac{1}{2}(p_1 + p_2). \tag{9.86}$$

Next, we rotate n clockwise 90° using the pseudoscalar e_{12} using

$$\hat{v} = \frac{e_{12}n}{|n|}$$

$$\hat{v} = \frac{e_{12}(p_2 - p_1)}{|p_2 - p_1|}. \tag{9.87}$$

Let's test (9.87) with the example shown in Figure 9.16.

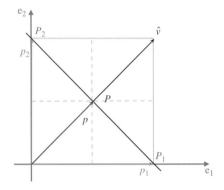

FIGURE 9.16.

Given

$$P_1 = (1, 0)$$
$$P_2 = (0, 1)$$
$$p_1 = e_1$$
$$p_2 = e_2$$
$$|p_2 - p_1| = \sqrt{2}. \tag{9.88}$$

Then, using (9.87)

$$\hat{v} = \frac{e_{12}(p_2 - p_1)}{|p_2 - p_1|}$$
$$\hat{v} = \frac{e_{12}(e_2 - e_1)}{\sqrt{2}} \tag{9.89}$$

and using (9.86)

$$p = \frac{1}{2}(p_1 + p_2)$$
$$= \frac{1}{2}(e_1 + e_2)$$
$$P = \left(\tfrac{1}{2}, \tfrac{1}{2}\right) \tag{9.90}$$

which are confirmed by the figure.

9.11 Intersection of two 2D lines

Let's explore how geometric algebra provides an alternative way of determining the intersection point of two 2D lines. We begin by defining two parametric lines as follows:

$$p = r + \lambda a$$
$$q = s + \varepsilon b \tag{9.91}$$

where

> r, s are position vectors for points R, S
>
> a, b are direction vectors for the lines
>
> p, q are position vectors for points P, Q on the lines, and
>
> λ, ε are scalar parameters for the lines

as shown in Figure 9.17.

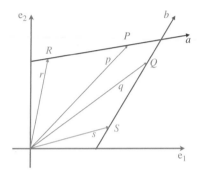

<small>Figure 9.17.</small>

The traditional algebraic approach is to equate (9.91) with (9.92):

$$r + \lambda a = s + \varepsilon b \tag{9.92}$$

which, after some manipulation, reveals that

$$\lambda = \frac{x_b(y_r - y_s) - y_b(x_r - x_s)}{x_a y_b - x_b y_a} \tag{9.93}$$

and permits us to calculate p using (9.91). Further algebraic manipulation reveals that

$$p = \frac{\begin{vmatrix} x_p & y_p \\ x_b & y_b \end{vmatrix}}{\begin{vmatrix} x_a & y_a \\ x_b & y_b \end{vmatrix}} a + \frac{\begin{vmatrix} x_p & y_p \\ x_a & y_a \end{vmatrix}}{\begin{vmatrix} x_b & y_b \\ x_a & y_a \end{vmatrix}} b. \tag{9.94}$$

Although (9.94) is highly symmetric, it actually references the point we are trying to find!

However, it just so happens that (9.94) is very close to the geometric algebra solution, that we will now examine.

We start again with the original line equations (9.91) but redraw Figure 9.17 in a slightly different way, as shown in Figure 9.18.

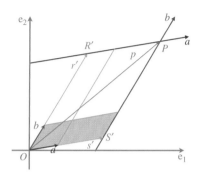

<small>FIGURE 9.18.</small>

The points R and S have been moved to R' and S' respectively to form a parallelogram $OS'PR'$ using the vectors a and b. It should be clear that s' is a multiple of a:

$$s' = \frac{s' \wedge b}{a \wedge b} a \tag{9.95}$$

and r' is a multiple of b:

$$r' = \frac{r' \wedge a}{b \wedge a} b. \tag{9.96}$$

The reason for this is that

$$\frac{s' \wedge b}{a \wedge b} \tag{9.97}$$

represents the ratio of two areas formed by two outer products sharing the common vector b, and

$$\frac{r' \wedge a}{b \wedge a} \tag{9.98}$$

represents the ratio of two areas formed by two outer products sharing the common vector a.

From Figure 9.18 we observe that

$$p = s' + r' \tag{9.99}$$

therefore,

$$p = \frac{s' \wedge b}{a \wedge b} a + \frac{r' \wedge a}{b \wedge a} b. \tag{9.100}$$

We now have the point of intersection defined in terms of two unknown points R' and R'.

But, fortunately, the nature of the outer product reminds us that these points are not important. For if we move R' and S' back to R and S, then

$$s \wedge b = s' \wedge b \tag{9.101}$$

and

$$r \wedge b = r' \wedge b \tag{9.102}$$

therefore,

$$p = \frac{s \wedge b}{a \wedge b} a + \frac{r \wedge a}{b \wedge a} b. \tag{9.103}$$

Let's put (9.103) to the test with an example.

Figure 9.19 shows two lines with equations:

$$p = r + \lambda a$$
$$p = s + \varepsilon b \qquad (9.104)$$

where

$$a = e_1 - e_2$$
$$b = e_1 + e_2$$
$$r = e_1 + 4e_2$$
$$s = e_1 + 2e_2 \qquad (9.105)$$

therefore, using (9.103)

$$
\begin{aligned}
p &= \frac{s \wedge b}{a \wedge b}a + \frac{r \wedge a}{b \wedge a}b \\
&= \frac{(e_1 + 2e_2) \wedge (e_1 + e_2)}{(e_1 - e_2) \wedge (e_1 + e_2)}(e_1 - e_2) + \frac{(e_1 + 4e_2) \wedge (e_1 - e_2)}{(e_1 + e_2) \wedge (e_1 - e_2)}(e_1 + e_2) \\
&= \frac{-e_{12}}{2e_{12}}(e_1 - e_2) + \frac{-5e_{12}}{-2e_{12}}(e_1 + e_2) \\
&= -\frac{1}{2}(e_1 - e_2) + \frac{5}{2}(e_1 + e_2) \\
p &= 2e_1 + 3e_2. \qquad (9.106)
\end{aligned}
$$

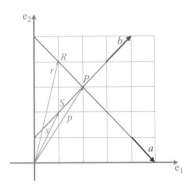

FIGURE 9.19.

This shows that we can rewrite (9.94) as

$$
p = \frac{\begin{vmatrix} x_s & y_s \\ x_b & y_b \end{vmatrix}}{\begin{vmatrix} x_a & y_a \\ x_b & y_b \end{vmatrix}}a + \frac{\begin{vmatrix} x_r & y_r \\ x_a & y_a \end{vmatrix}}{\begin{vmatrix} x_b & y_b \\ x_a & y_a \end{vmatrix}}b \qquad (9.107)
$$

and substituting the values we have

$$
p = \frac{\begin{vmatrix} 1 & 2 \\ 1 & 1 \end{vmatrix}}{\begin{vmatrix} 1 & -1 \\ 1 & 1 \end{vmatrix}}(e_1 - e_2) + \frac{\begin{vmatrix} 1 & 4 \\ 1 & -1 \end{vmatrix}}{\begin{vmatrix} 1 & 1 \\ 1 & -1 \end{vmatrix}}(e_1 + e_2)
$$

$$
= \frac{-1}{2}(e_1 - e_2) + \frac{-5}{-2}(e_1 + e_2)
$$

$$
p = 2e_1 + 3e_2. \tag{9.108}
$$

9.12 Intersection of two 2D lines using homogeneous coordinates

Now let's explore a completely different approach to the line intersection problem using homogeneous coordinates.

By embedding our 2D scenario in a space with an extra dimension, we create the opportunity of describing line segments as a blade - which, in turn, permits us to use the meet operation. Figure 9.20 shows this scenario.

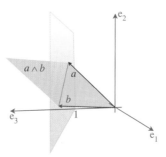

FIGURE 9.20.

The blade $a \wedge b$ has a unique description where a and b have homogeneous coordinates:

$$
a = a_1 e_1 + a_2 e_2 + e_3
$$

$$
b = b_1 e_1 + b_2 e_2 + e_3. \tag{9.109}
$$

Given two such blades: $a \wedge b$ and $c \wedge d$, we can find their homogeneous intersection using the meet operation:

$$
(a \wedge b) \vee (c \wedge d) = (a \wedge b)^* \cdot (c \wedge d). \tag{9.110}
$$

Figure 9.21 shows the lines from the previous example, where

$$a = e_1 + 4e_2 + e_3$$
$$b = 5e_1 + e_3$$
$$c = e_2 + e_3$$
$$d = 3e_1 + 4e_2 + e_3. \tag{9.111}$$

Therefore,

$$a \wedge b = (e_1 + 4e_2 + e_3) \wedge (5e_1 + e_3)$$
$$= e_{13} + 20e_{21} + 4e_{23} + 5e_{31}$$
$$a \wedge b = 20e_{21} + 4e_{23} + 4e_{31}$$
$$(a \wedge b)^* = e_{123}(20e_{21} + 4e_{23} + 4e_{31})$$
$$= 20e_{12321} + 4e_{12323} + 4e_{12331}$$
$$(a \wedge b)^* = 20e_3 - 4e_1 - 4e_2$$
$$c \wedge d = (e_2 + e_3) \wedge (3e_1 + 4e_2 + e_3)$$
$$= 3e_{21} + e_{23} + 3e_{31} + 4e_{32}$$
$$c \wedge d = 3e_{21} + 3e_{32} + 3e_{31} \tag{9.112}$$

and

$$(a \wedge b)^* \cdot (c \wedge b) = (20e_3 - 4e_1 - 4e_2) \cdot (3e_{21} + 3e_{32} + 3e_{31})$$
$$= 60e_2 + 60e_1 + 12e_2 + 12e_3 - 12e_1 + 12e_3$$
$$= 48e_1 + 72e_2 + 24e_3. \tag{9.113}$$

As 24 is the homogeneous scaling factor, this makes

$$(a \wedge b)^* \cdot (c \wedge d) = 2e_1 + 3e_2 \tag{9.114}$$

and confirms the previous result.

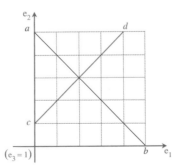

FIGURE 9.21.

9.13 Orientation of a point and a 2D line

Given an oriented 2D line segment AB, a point P can be to the left, to the right, or on the line. These relative orientations are revealed by the outer product $\overrightarrow{AB} \wedge \overrightarrow{AP}$:

If

$$\overrightarrow{AB} \wedge \overrightarrow{AP} \text{ is } +ve, P \text{ is to the left.}$$

$$\overrightarrow{AB} \wedge \overrightarrow{AP} \text{ is } -ve, P \text{ is to the right.}$$

$$\overrightarrow{AB} \wedge \overrightarrow{AP} \text{ is zero, } P \text{ is on the line.}$$

Figure 9.22 shows a line segment $\overrightarrow{AB}$ and three test points P, Q and R.

Therefore, given

$$\overrightarrow{AB} = 3e_1 + 3e_2$$

$$\overrightarrow{AP} = e_1 + 3e_2$$

$$\overrightarrow{AQ} = 3e_1$$

$$\overrightarrow{AR} = e_1 + e_2 \tag{9.115}$$

then

$$\overrightarrow{AB} \wedge \overrightarrow{AP} = (3e_1 + 3e_2) \wedge (e_1 + 3e_2)$$

$$= 9e_{12} - 3e_{12}$$

$$= 6e_{12} \text{ (i.e. to the left)}$$

$$\overrightarrow{AB} \wedge \overrightarrow{AQ} = (3e_1 + 3e_2) \wedge 3e_1$$

$$= -9e_{12} \text{ (i.e. to the right)}$$

$$\overrightarrow{AB} \wedge \overrightarrow{AR} = (3e_1 + 3e_2) \wedge (e_1 + e_2)$$

$$= 3e_{12} - 3e_{12}$$

$$= 0 \text{ (i.e. on the line).} \tag{9.116}$$

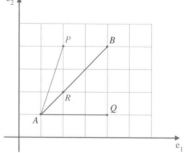

Figure 9.22.

9.14 Intersection of a line and a circle

We now consider the problem of calculating the intersection between a line and a circle. Figure 9.23 shows a typical scenario where a line represented by

$$p = t + \lambda \hat{v} \tag{9.117}$$

intersects a circle of radius r and centre C with position vector

$$c = x_c e_1 + y_c e_2. \tag{9.118}$$

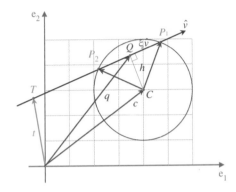

FIGURE 9.23.

The strategy is to establish a point Q on the line such that CQ is perpendicular to $\hat{v}$. If $\overrightarrow{QP_1} = \zeta \hat{v}$, then as $|\overrightarrow{CP_1}| = r$ we can find ζ using the theorem of Pythagoras. We begin with the geometric product

$$q\hat{v} = q \cdot \hat{v} + q \wedge \hat{v} \tag{9.119}$$

but

$$q \cdot \hat{v} = c \cdot \hat{v} \tag{9.120}$$

and

$$q \wedge \hat{v} = t \wedge \hat{v} \tag{9.121}$$

therefore

$$q\hat{v} = c \cdot \hat{v} + t \wedge \hat{v}. \tag{9.122}$$

Dividing throughout by $\hat{v}$ we have

$$q = (c \cdot \hat{v} + t \wedge \hat{v})\hat{v}. \tag{9.123}$$

From Figure 9.23

$$\overrightarrow{CQ} = q - c \tag{9.124}$$

and

$$|\overrightarrow{CQ}| = |q - c| = h. \tag{9.125}$$

From $\triangle QCP_1$

$$\zeta = \sqrt{r^2 - h^2} \tag{9.126}$$

and

$$p_1 = c + \overrightarrow{CQ} + \zeta\hat{v} \tag{9.127}$$

and

$$p_2 = c + \overrightarrow{CQ} - \zeta\hat{v}. \tag{9.128}$$

If $P_1 = P_2$ the line is tangent to the circle.

Let's test (9.128) with the example shown in Figure 9.24 where

$$
\begin{aligned}
T &= (0, 2) & t &= 2e_2 \\
C &= (3, 3) & c &= 3e_1 + 3e_2 \\
r &= 2 & \hat{v} &= \frac{1}{\sqrt{2}}(e_1 + e_2).
\end{aligned} \tag{9.129}
$$

Using (9.123) we have

$$
\begin{aligned}
q &= \left(c \cdot \hat{v} + t \wedge \hat{v}\right)\hat{v} \\
&= \left[(3e_1 + 3e_2) \cdot \frac{1}{\sqrt{2}}(e_1 + e_2) + 2e_2 \wedge \frac{1}{\sqrt{2}}(e_1 + e_2)\right]\frac{1}{\sqrt{2}}(e_1 + e_2) \\
&= \left[\frac{6}{\sqrt{2}} + \frac{2}{\sqrt{2}}e_{21}\right]\frac{1}{\sqrt{2}}(e_1 + e_2) \\
&= \frac{1}{2}(6e_1 + 6e_2 + 2e_2 - 2e_1) \\
q &= 2e_1 + 4e_2. \tag{9.130}
\end{aligned}
$$

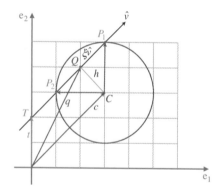

FIGURE 9.24.

Therefore using (9.124)

$$\overrightarrow{CQ} = 2e_1 + 4e_2 - 3e_1 - 3e_2 = -e_1 + e_2 \tag{9.131}$$

and

$$h = \sqrt{2}. \tag{9.132}$$

Therefore, using (9.126) we have

$$\zeta = \sqrt{4 - 2} = \sqrt{2} \tag{9.133}$$

and

$$p_1 = c + \overrightarrow{CQ} + \zeta \hat{v}$$

$$= (3e_1 + 3e_2) + (-e_1 + e_2) + \sqrt{2}\,\frac{1}{\sqrt{2}}\,(e_1 + e_2)$$

$$p_1 = 3e_1 + 5e_2. \tag{9.134}$$

Similarly, using (9.128)

$$p_2 = c + \overrightarrow{CQ} - \zeta \hat{v}$$

$$= (3e_1 + 3e_2) + (-e_1 + e_2) - \sqrt{2}\,\frac{1}{\sqrt{2}}\,(e_1 + e_2)$$

$$p_2 = e_1 + 3e_2. \tag{9.135}$$

9.15 Orientation of a point with a plane

Having shown that the outer product identifies the relative orientation of a point with a 2D line, we will now demonstrate that it also works for a plane. For example, if a bivector is used to represent the orientation of a plane, the outer product of the point's position vector with the bivector computes an oriented volume. Figure 9.25 shows a bivector $a \wedge b$ and a test point P with vector p relative to the bivector.

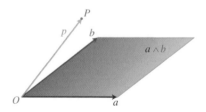

FIGURE 9.25.

If

$a \wedge b \wedge p$ is $+ve$, then p is 'above' the bivector.

$a \wedge b \wedge p$ is $-ve$, then p is 'below' the bivector.

$a \wedge b \wedge p$ is zero, then p is coplanar with the bivector.

The terms 'above' and 'below' mean in the bivector's positive and negative half-space respectively.

Figure 9.26 shows a plane whose orientation is represented by the bivector $a \wedge b$, and three test points P, Q and R.

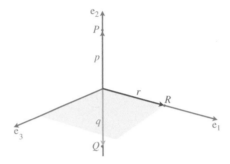

FIGURE 9.26.

If $P = (0, 1, 0)$ $Q = (0, -1, 0)$ $R = (1, 0, 0)$

$$a = e_1 + e_3$$
$$b = e_1 \tag{9.136}$$

then

$$p = e_2$$
$$q = -e_2$$
$$r = e_1 \tag{9.137}$$

and

$$a \wedge b \wedge p = (e_1 + e_3) \wedge e_1 \wedge e_2$$
$$= e_{123} \text{ (i.e. } P \text{ is 'above')}$$
$$a \wedge b \wedge q = (e_1 + e_3) \wedge e_1 \wedge (-e_2)$$
$$= -e_{123} \text{ (i.e. } Q \text{ is 'below')}$$
$$a \wedge b \wedge r = (e_1 + e_3) \wedge e_1 \wedge e_1$$
$$= 0 \text{ (i.e. } R \text{ is coplanar).} \tag{9.138}$$

9.16 Plane equation

Vector analysis provides a simple way to describe plane equations. For instance, (9.139) is often used to describe a plane

$$ax + by + cz = d \tag{9.139}$$

where

$$\hat{n} = ai + bj + ck \tag{9.140}$$

is a unit vector perpendicular to the plane, and d is the perpendicular distance from the plane to the origin. Furthermore, given any point P on the plane with position vector p, then

$$\hat{n} \cdot p = d. \tag{9.141}$$

Geometric algebra provides an alternative model using a bivector. For instance, if a unit bivector $\hat{B}$ provides the orientation of a plane, then its perpendicular vector $\hat{n}$ is obtained using

$$\hat{n} = -\hat{B}^* \tag{9.142}$$

which permits us to write the geometric algebra equivalent of (9.141) as

$$-\hat{B}^* \cdot p = d. \tag{9.143}$$

For example, if

$$\hat{B} = \frac{1}{\sqrt{3}}(e_{12} + e_{23} + e_{31}) \tag{9.144}$$

then

$$\hat{n} = -\hat{B}^*$$
$$= -\frac{1}{\sqrt{3}}(e_{12} + e_{23} + e_{31})e_{123}$$
$$\hat{n} = \frac{1}{\sqrt{3}}(e_1 + e_2 + e_3). \tag{9.145}$$

Now if we set the perpendicular distance to

$$d = \frac{\sqrt{3}}{3} \tag{9.146}$$

this identifies a plane that intersects the three axes at $(1, 0, 0)$, $(0, 1, 0)$, $(0, 0, 1)$ as shown in Figure 9.27.

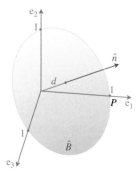

FIGURE 9.27.

Therefore, using (9.143)

$$-\hat{B}^* \cdot p = d$$

$$\frac{1}{\sqrt{3}}(e_1 + e_2 + e_3) \cdot p = \frac{\sqrt{3}}{3}$$

$$(e_1 + e_2 + e_3) \cdot p = 1 \tag{9.147}$$

and if $P = (1, 0, 0)$ then $p = e_1$ and (9.147) is upheld.

Let's consider an alternative model where a bivector B defines a plane'ts orientation and a point Q with position vector q is a known point on the plane. For another point P on the plane with position vector p

$$\langle (p - q)B \rangle_3 = 0. \tag{9.148}$$

For example, consider the scenario shown in Figure 9.28 where

$$B = e_{31}$$

$$Q = (0, 1, 0)$$

$$q = e_2. \tag{9.149}$$

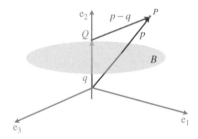

Figure 9.28.

If

$$P = (1, 2)$$

$$P = e_1 + 2e_2. \tag{9.150}$$

For P to be on the plane (which it is not)

$$\langle (p - q)B \rangle_3 = 0$$

$$= \langle (e_1 + 2e_2 - e_2)e_{31} \rangle_3$$

$$= \langle (e_1 + e_2)e_{31} \rangle_3$$

$$= \langle (-e_3 + e_{123}) \rangle_3$$

$$= 1 \tag{9.151}$$

therefore, P is 'above' the plane.

If, on the other hand

$$P = (1, 1)$$

$$p = e_1 + e_2 \tag{9.152}$$

for P to be on the plane (which it is)

$$\langle (p - q)B \rangle_3 = 0$$
$$= \langle (e_1 + e_2 - e_2)e_{31} \rangle_3$$
$$= \langle -e_3 \rangle_3$$
$$= 0 \tag{9.153}$$

therefore, P is on the plane.
   Finally, if

$$P = (1, 0)$$

$$p = e_1 \tag{9.154}$$

for P to be on the plane (which it is not)

$$\langle (p - q)B \rangle_3 = 0$$
$$= \langle (e_1 - e_2)e_{31} \rangle_3$$
$$= \langle -e_3 - e_{123} \rangle_3$$
$$= -1 \tag{9.155}$$

therefore, P is 'below' the plane.

9.17 Orientation of a point with a convex object

We have just seen that a bivector provides a useful geometric entity for partitioning space. Therefore, given access to a bivector B and a point Q on the bivector's plane, we will develop a technique that permits us to test whether a point is inside, touching or outside a convex object.

Figure 9.29 shows a plane whose orientation is determined by a bivector B. Q is a point on the plane with position vector q, and P is a test point, not necessarily on the plane, with position vector p. If we form the product $(p - q)B$ two things happen: the inner product rotates the component of $(p - q)$ coplanar with B 90°, and the outer product creates a volume from the component of $(p - q)$ perpendicular to B. Consequently, the position of P is determined by $(p - q) \wedge B$ or its equivalent $\langle (p - q)B \rangle_3$:

$$\langle (p - q)B \rangle_3 = +ve, \ P \text{ is 'above' the plane}$$
$$\langle (p - q)B \rangle_3 = -ve, \ P \text{ is 'below' the plane}$$
$$\langle (p - q)B \rangle_3 = 0, \ P \text{ is 'touching' the plane.} \tag{9.156}$$

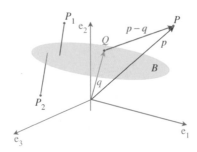

Figure 9.29.

This mechanism can be used to determine whether a point is inside, outside or on the surface of a convex polyhedron. To illustrate the technique we construct a unit cube as shown in Figure 9.30. The figure gives six views of the cube where each side is identified in turn with its bivector orientation, direction for outside, a point Q on the plane and its position vector q.

These data are summarized in Table 9.1.

Table 9.1   Data associated with the six sides of a unit cube

Side	Bivector	Q	q
1	$-e_{12}$	$(1, 0, 0)$	e_1
2	e_{23}	$(1, 0, 1)$	$e_1 + e_3$
3	e_{12}	$(0, 0, 1)$	e_3
4	$-e_{23}$	$(0, 1, 0)$	e_2
5	e_{31}	$(0, 1, 0)$	e_2
6	$-e_{31}$	$(1, 0, 1)$	$e_1 + e_3$

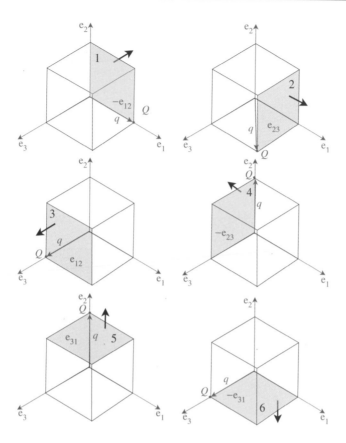

Figure 9.30.

We can now test to see whether any other point P is inside, outside or on the cube's surface. Let's begin with $P = (0, 2, 0)$, which is clearly outside the cube's boundary. We investigate each side using (9.156) and show the results in Table 9.2.

Table 9.2 records that P potentially touches sides 1 and 4; is potentially 'inside' sides 2, 3 and 6 and outside side 5. To be inside, every test must return an inside result. To be outside, only one test need be valid. Consequently, P is classified as outside the cuboid.

Next, we test another point $P = \left(\frac{1}{2}, \frac{1}{2}, \frac{1}{2}\right)$, which is clearly inside the cube.

Table 9.3 records that all the results are negative, and confirms that P is inside the enclosed volume.

Finally, we test another point $P = (1, 1, 1)$, which touches the cube. Table 9.4 records that sides 2, 3 and 5 touch the point P, which can only be at a vertex. The other sides record the point being potentially inside. Consequently, point P touches the cube where sides 2, 3 and 5 intersect.

TABLE 9.2   Testing each side of a cube with (9.156)

Side	$(p-q)B$	$\langle (p-q)B \rangle_3$
1	$-(-e_1 + 2e_2)e_{12}$	0
2	$(2e_2 - e_1 - e_3)e_{23}$	-1
3	$(2e_2 - e_3)e_{12}$	-1
4	$-(2e_2 - e_2)e_{23}$	0
5	$(2e_2 - e_2)e_{31}$	1
6	$-(2e_2 - e_1 - e_3)e_{31}$	-2

TABLE 9.3   Testing each side of a cube with (9.156)

Side	$(p-q)B$	$\langle (p-q)B \rangle_3$
1	$-(-\frac{1}{2}e_1 + \frac{1}{2}e_2 + \frac{1}{2}e_3)e_{12}$	$-\frac{1}{2}$
2	$(-\frac{1}{2}e_1 + \frac{1}{2}e_2 - \frac{1}{2}e_3)e_{23}$	$-\frac{1}{2}$
3	$(\frac{1}{2}e_1 + \frac{1}{2}e_2 - \frac{1}{2}e_3)e_{12}$	$-\frac{1}{2}$
4	$-(\frac{1}{2}e_1 - \frac{1}{2}e_2 + \frac{1}{2}e_3)e_{23}$	$-\frac{1}{2}$
5	$(\frac{1}{2}e_1 - \frac{1}{2}e_2 + \frac{1}{2}e_3)e_{31}$	$-\frac{1}{2}$
6	$-(-\frac{1}{2}e_1 + \frac{1}{2}e_2 - \frac{1}{2}e_3)e_{31}$	$-\frac{1}{2}$

TABLE 9.4   Testing each side of a cube with (9.156)

Side	$(p-q)B$	$\langle (p-q)B \rangle_3$
1	$-(e_2 + e_3)e_{12}$	-1
2	$(e_2)e_{23}$	0
3	$(e_1 + e_2)e_{12}$	0
4	$-(e_1 + e_3)e_{23}$	-1
5	$(e_1 + e_3)e_{31}$	0
6	$-(e_2)e_{31}$	-1

9.18 Angle between a vector and a bivector

As a bivector defines the orientation of a plane, it is relatively easy to determine the angle between it and a vector - we simply take the dual of the bivector to create the perpendicular axial vector and use the inner product to reveal the inner angle. Figure 9.31 shows a scenario where bivector B has a normal n where

$$n = -B^*. \tag{9.157}$$

The minus sign is introduced to establish the convention that the axial vector is located in the bivector's positive half-space.

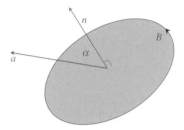

FIGURE 9.31.

α is found using the inner product:

$$\alpha = \cos^{-1}\left(\frac{a \cdot n}{|a||n|}\right) = \cos^{-1}\left(\frac{-a \cdot B^*}{|a||n|}\right). \tag{9.158}$$

For example, consider the following conditions:

$$B = e_{12}$$
$$a = -e_1 + e_3. \tag{9.159}$$

Therefore, using (9.158)

$$n = -e_{123}B$$
$$= -e_{123}e_{12}$$
$$n = e_3$$

and

$$\alpha = \cos^{-1}\left(\frac{(-e_1 + e_3) \cdot e_3}{\sqrt{2} \times 1}\right)$$
$$= \cos^{-1}\left(\frac{1}{\sqrt{2}}\right)$$
$$\alpha = 45°. \tag{9.160}$$

9.19 Angle between two bivectors

We can find the angle between two bivectors by first taking their duals - which are perpendicular vectors - and then use the inner product to reveal their separating angle. Figure 9.32 shows a typical scenario with bivectors B and C, and their respective duals m and n.

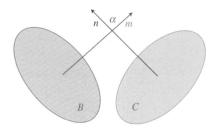

FIGURE 9.32.

Therefore,

$$\alpha = \cos^{-1}\left(\frac{m \cdot n}{|m||n|}\right) = \cos^{-1}\left(\frac{B^* \cdot C^*}{|B^*||C^*|}\right). \tag{9.161}$$

For example, consider the two bivectors in Figure 9.33, where $B = a \wedge b$ and $C = c \wedge d$.
Given

$$a = e_1$$
$$b = e_2 - e_3$$
$$c = -e_1 + e_2$$
$$d = e_3 \tag{9.162}$$

then

$$a \wedge b = e_1 \wedge (e_2 - e_3)$$
$$a \wedge b = e_{12} + e_{31}$$
$$c \wedge d = (-e_1 + e_2) \wedge e_3$$
$$c \wedge d = e_{23} + e_{31}. \tag{9.163}$$

Therefore,

$$m = B^* = e_{123}(e_{12} + e_{31})$$
$$m = -e_3 - e_2 \tag{9.164}$$

and

$$n = C^* = e_{123}(e_{23} + e_{31})$$
$$n = -e_1 - e_2. \tag{9.165}$$

Therefore,

$$\alpha = \cos^{-1}\left(\frac{(-e_2 - e_3)\cdot(-e_1 - e_2)}{\sqrt{2}\sqrt{2}}\right)$$

$$= \cos^{-1}\left(\frac{1}{2}\right)$$

$$\alpha = 60°. \tag{9.166}$$

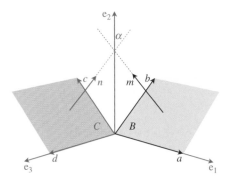

FIGURE 9.33.

9.20 Intersection of a line and a plane

A standard approach to find the intersection between a line and a plane is to substitute the parametric line equation into the plane equation. For instance, if the line equation is

$$p = t + \lambda v \tag{9.167}$$

where

t is the position vector for a point T on the line

v is the line's direction vector

λ is a scalar

p is the position vector of a point P on the line

and the plane equation is

$$ax + by + cz = \delta \tag{9.168}$$

where

$n = ai + bj + ck$ is the plane's normal vector

δ is the perpendicular distance from the plane to the origin then the point of intersection is

$$p = t + \left(\frac{\delta - n \cdot t}{n \cdot v} \right) v. \tag{9.169}$$

On the other hand, if the plane is defined in terms of its normal vector n and a known point Q with position vector q, then the point of intersection is

$$p = t + \left[\frac{(q - t) \cdot n}{v \cdot n} \right] v. \tag{9.170}$$

However, in geometric algebra we would normally employ a bivector to describe a plane's orientation, which would transform (9.170) into

$$p = t + \left[\frac{(q - t) \cdot IB}{v \cdot IB} \right] v \tag{9.171}$$

where

I is the pseudoscalar

and

B is the plane's bivector.

A slightly different approach is shown in Figure 9.34 where we see a plane defined by a bivector B and a point Q on the plane. The line equation is the same as (9.167).

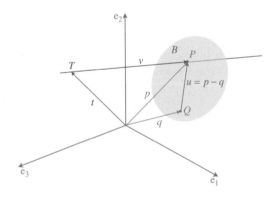

FIGURE 9.34.

From the figure

$$u = p - q \tag{9.172}$$

and

$$B \wedge u = 0 \tag{9.173}$$

therefore,

$$B \wedge (p - q) = 0. \tag{9.174}$$

Substituting the line equation (9.167) in (9.174) we have

$$B \wedge (t + \lambda v - q) = 0$$

$$B \wedge (t - q) + B \wedge \lambda v = 0 \tag{9.175}$$

therefore,

$$\lambda = \frac{B \wedge (q - t)}{B \wedge v} \tag{9.176}$$

and

$$p = t + \left(\frac{B \wedge (q - t)}{B \wedge v} \right) v. \tag{9.177}$$

Let's test (9.177) with the scenario shown in Figure 9.35 where

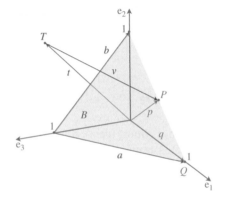

FIGURE 9.35.

$$q = e_1$$
$$a = e_1 - e_3$$
$$b = e_2 - e_3 \tag{9.178}$$

then

$$B = a \wedge b$$
$$= (e_1 - e_3) \wedge (e_2 - e_3)$$
$$B = e_{12} + e_{23} + e_{31}.$$
$$T = (0, 1, 1) \text{ and } t = e_2 + e_3$$
$$Q = (1, 0, 0) \text{ and } q = e_1 \tag{9.179}$$

and the line's direction vector is

$$v = \tfrac{1}{2} e_1 - \tfrac{1}{2} e_2 - e_3. \tag{9.180}$$

Therefore, using (9.177)

$$p = t + \left(\frac{B \wedge (q - t)}{B \wedge v}\right) v.$$

$$= (e_2 + e_3) + \left(\frac{(e_{12} + e_{23} + e_{31}) \wedge (e_1 - e_2 - e_3)}{(e_{12} + e_{23} + e_{31}) \wedge (\frac{1}{2}e_1 - \frac{1}{2}e_2 - e_3)}\right) \left(\frac{1}{2}e_1 - \frac{1}{2}e_2 - e_3\right)$$

$$= (e_2 + e_3) + \left(\frac{e_{123} - e_{132} + e_{321}}{e_{123} - \frac{1}{2}e_{132} + \frac{1}{2}e_{321}}\right) \left(\frac{1}{2}e_1 - \frac{1}{2}e_2 - e_3\right)$$

$$= (e_2 + e_3) + \left(\frac{1}{2}e_1 - \frac{1}{2}e_2 - e_3\right)$$

$$p = \frac{1}{2}e_1 + \frac{1}{2}e_2 \tag{9.181}$$

which makes

$$Q = \left(\frac{1}{2}, \frac{1}{2}, 0\right). \tag{9.182}$$

9.21 Intersection of a line and a sphere

The line-sphere intersection problem is almost identical to the line-circle intersection problem, however, for completeness it is repeated again.

Figure 9.36 shows a scenario where a line

$$p = t + \lambda \hat{v} \tag{9.183}$$

intersects a sphere of radius r and centre C with position vector

$$c = x_c e_1 + y_c e_2 + z_c e_3. \tag{9.184}$$

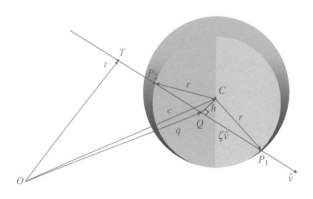

FIGURE 9.36.

The strategy is to establish a point Q on the line such that CQ is perpendicular to $\hat{v}$. If $\overrightarrow{QP_1} = \zeta\hat{v}$, then as $|\overrightarrow{CP_1}| = r$ we can find ζ using the theorem of Pythagoras. We begin with the geometric product

$$q\hat{v} = q \cdot \hat{v} + q \wedge \hat{v} \qquad (9.185)$$

but

$$q \cdot \hat{v} = c \cdot \hat{v} \qquad (9.186)$$

and

$$q \wedge \hat{v} = t \wedge \hat{v} \qquad (9.187)$$

therefore

$$q\hat{v} = c \cdot \hat{v} + t \wedge \hat{v}. \qquad (9.188)$$

Dividing throughout $\hat{v}$ by we have

$$q = (c \cdot \hat{v} + t \wedge \hat{v})\hat{v}. \qquad (9.189)$$

From Figure 9.36

$$\overrightarrow{CQ} = q - c \qquad (9.190)$$

and

$$|\overrightarrow{CQ}| = |q - c| = h. \qquad (9.191)$$

From $\triangle QCP_1$

$$\zeta = \sqrt{r^2 - h^2} \qquad (9.192)$$

and

$$p_1 = c + \overrightarrow{CQ} + \zeta\hat{v} \qquad (9.193)$$

and

$$p_2 = c + \overrightarrow{CQ} - \zeta\hat{v}. \qquad (9.194)$$

If $P_1 = P_2$ the line is tangent to the sphere.

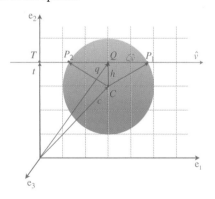

FIGURE 9.37.

Let's test (9.193) with the example shown in Figure 9.37 where

$$T = (0, 4) \quad t = 4e_2$$
$$C = (3, 3) \quad c = 3e_1 + 3e_2$$
$$r = 2 \qquad \hat{v} = e_1. \tag{9.195}$$

Using (9.189) we have

$$q = (c \cdot \hat{v} + t \wedge \hat{v})\hat{v}$$
$$= [(3e_1 + 3e_2) \cdot e_1 + 4e_2 \wedge e_1] e_1$$
$$= (3 + 4e_{21})e_1$$
$$q = 3e_1 + 4e_2. \tag{9.196}$$

Therefore using (9.190)

$$\overrightarrow{CQ} = 3e_1 + 4e_2 - 3e_1 - 3e_2 = e_2 \tag{9.197}$$

and

$$h = 1. \tag{9.198}$$

Therefore, using (9.192) we have

$$\zeta = \sqrt{4 - 1} = \sqrt{3} \tag{9.199}$$

and

$$p_1 = c + \overrightarrow{CQ} + \zeta \hat{v}$$
$$= (3e_1 + 3e_2) + e_2 + \sqrt{3}e_1$$
$$p_1 = (3 + \sqrt{3})e_1 + 4e_2. \tag{9.200}$$

Similarly, using (9.194)

$$p_2 = c + \overrightarrow{CQ} - \zeta \hat{v}$$
$$= (3e_1 + 3e_2) + e_2 - \sqrt{3}e_1$$
$$p_2 = (3 - \sqrt{3})e_1 + 4e_2. \tag{9.201}$$

9.22　Ray intersection and reflection

Figure 9.38 shows a scenario where a ray with line equation

$$p = t + \lambda v \tag{9.202}$$

is incident to a surface with orientation B and containing a known point Q. The point of intersection is given by

$$p = t + \left(\frac{B \wedge (q - t)}{B \wedge v}\right) v \tag{9.203}$$

therefore, the equation of the reflected ray is

$$p_R = p - \zeta \hat{B} v \hat{B} \qquad\qquad \text{where } [\zeta \in \mathbb{R}] \qquad\qquad (9.204)$$

which is rather compact.

In order to keep the following algebraic expansions simple, let's make $\hat{B} = e_{12}$ and $q = 0$, which simplifies (9.203) to

$$p = t - \left(\frac{B \wedge t}{B \wedge v} \right) v. \qquad\qquad (9.205)$$

Given

$$\cdot \; t = e_3$$

$$v = e_1 + e_2 - e_3$$

$$\hat{B} = e_{12}$$

$$\hat{B}^* = -e_3 \qquad\qquad (9.206)$$

then

$$p = e_3 - \left(\frac{e_{12} \wedge e_3}{e_{12} \wedge (e_1 + e_2 - e_3)} \right) (e_1 + e_2 - e_3)$$

$$= e_3 - \left(\frac{e_{123}}{-e_{123}} \right) (e_1 + e_2 - e_3)$$

$$p = e_1 + e_2 \qquad\qquad (9.207)$$

which is the point of intersection.

Next, using (9.204) the reflected ray is

$$v_R = -e_3(e_1 + e_2 - e_3)e_3 \qquad\qquad (9.208)$$

and

$$p_R = (e_1 + e_2) + \zeta v_R$$

$$= (e_1 + e_2) - \zeta e_3(e_1 + e_2 - e_3)e_3$$

$$p_R = e_1 + e_2 + \zeta(e_1 + e_2 + e_3). \qquad\qquad (9.209)$$

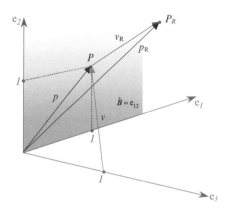

Figure 9.38.

9.23 Summary

The above examples illustrate how the inner, outer and geometric products resolve a variety of geometric problems. The only reason I have given a step-by-step record of the expansions is to provide an insight into the workings of geometric algebra. However, one must bear in mind that software would normally undertake these product expansions.

What is really interesting is that all the problems have been resolved without recourse to the cross-product. But you have probably realised that it is still accessible by taking the negative dual of a bivector.

10 Conclusion

If you have arrived at this concluding chapter, having understood the previous nine chapters, then you will have understood the principles of geometric algebra. What I have tried to demonstrate is that vectorial quantities can be manipulated by an algebra that is similar to the algebra associated with real quantities. An algebra where one can add, subtract, multiply and even divide such vectorial quantities, so long as we obey the associated axioms. Fortunately, these axioms are not too onerous and are readily committed to memory.

Geometric algebra introduces a number of new concepts that can initially take some getting used to, especially if one has been educated along the lines of traditional vector analysis. But a flexible brain can easily be remolded into accepting new concepts such as bivectors, trivectors, pseudoscalars, multivectors, non-commutativity, etc., without too much effort.

Having shown the rudimentary action of geometric algebra, some readers may wish to pursue the subject to higher levels, especially in the form of vector subspaces. Others may wish to explore conformal geometric algebra, which promises even greater rewards; however, there is a US patent that covers its commercial exploitation.

An important issue for geometric algebra is what software system should be used? Well, as I mentioned in my previous book *Geometric Algebra for Computer Graphics*, systems such as Maple, Mathematica and MatLab all offer tools for manipulating the structures encountered in geometric algebra. At a personal level, one could implement a general multivector product algorithm where a pair of multivectors are multiplied to reveal a new multivector. However, this involves 1 scalar, 3 vectors, 3 bivectors and 1 pseudoscalar, which is rather top heavy. On the other hand, there is a strong case for implementing discrete functions for each of the important products employed in the previous chapters.

When I first stumbled across geometric algebra, I was astonished that such an algebra could exist without my knowledge. But having come to terms with its axioms and concepts, my view of mathematics has been transformed out of all recognition. Hopefully, I have been able to communicate this knowledge to you and change your own understanding of mathematics.

J. Vince, *Geometric Algebra: An Algebraic System for Computer Games and Animation*,
© Springer-Verlag London Limited 2009

Appendix A

Proof. The outer product is associative

Given three vectors

$$a = a_1 e_1 + a_2 e_2 + a_3 e_3$$
$$b = b_1 e_1 + b_2 e_2 + b_3 e_3$$
$$c = c_1 e_1 + c_2 e_2 + c_3 e_3 \tag{A.1}$$

and

$$
\begin{aligned}
b \wedge c &= (b_1 e_1 + b_2 e_2 + b_3 e_3) \wedge (c_1 e_1 + c_2 e_2 + c_3 e_3) \\
&= b_1 c_2 e_1 \wedge e_2 + b_1 c_3 e_1 \wedge e_3 \\
&\quad + b_2 c_1 e_2 \wedge e_1 + b_2 c_3 e_2 \wedge e_3 \\
&\quad + b_3 c_1 e_3 \wedge e_1 + b_3 c_2 e_3 \wedge e_2 \\
&= (b_1 c_2 - b_2 c_1) e_1 \wedge e_2 + (b_2 c_3 - b_3 c_2) e_2 \wedge e_3 + (b_3 c_1 - b_1 c_3) e_3 \wedge e_1.
\end{aligned} \tag{A.2}
$$

Therefore,

$$
\begin{aligned}
a \wedge b \wedge c &= (a_1 e_1 + a_2 e_2 + a_3 e_3) \wedge (b \wedge c) \\
&= a_1 (b_2 c_3 - b_3 c_2) e_{123} + a_2 (b_3 c_1 - b_1 c_3) e_{231} + a_3 (b_1 c_2 - b_2 c_1) e_{312} \\
&= (a_1 b_2 c_3 - a_1 b_3 c_2) e_{123} + (a_2 b_3 c_1 - a_2 b_1 c_3) e_{123} + (a_3 b_1 c_2 - a_3 b_2 c_1) e_{123}
\end{aligned} \tag{A.3}
$$

$$
a \wedge (b \wedge c) = \begin{vmatrix} a_1 & b_1 & c_1 \\ a_2 & b_2 & c_2 \\ a_3 & b_3 & c_3 \end{vmatrix}. \tag{A.4}
$$

J. Vince, *Geometric Algebra: An Algebraic System for Computer Games and Animation*,
© Springer-Verlag London Limited 2009

Similarly, it can be shown that

$$(a \wedge b) \wedge c = \begin{vmatrix} a_1 & b_1 & c_1 \\ a_2 & b_2 & c_2 \\ a_3 & b_3 & c_3 \end{vmatrix} \tag{A.5}$$

therefore, the outer product is associative. $\blacksquare$

Appendix B

In this appendix we show the full expansion of the outer product of three vectors.

Given three vectors

$$a = a_1 e_1 + a_2 e_2 + a_3 e_3$$
$$b = b_1 e_1 + b_2 e_2 + b_3 e_3 \quad \text{(B.1)}$$
$$c = c_1 e_1 + c_2 e_2 + c_3 e_3$$

therefore, using (4.35)

$$a \wedge b \wedge c = \frac{1}{3!} \left[(abc - cba) + (bca - acb) + (cab - bac) \right] \quad \text{(B.2)}$$

or in determinant form

$$a \wedge b \wedge c = \frac{1}{3!} \begin{vmatrix} a & a & a \\ b & b & b \\ c & c & c \end{vmatrix}. \quad \text{(B.3)}$$

Table B.1 shows the six permutations of the geometric products:

$$abc, cba, bca, acb, cab, bac. \quad \text{(B.4)}$$

But as explained in chapter 3, the odd permutations must have their signs switched. These switched sign products are shown in Table B.2, together with the sum of each row.

One can see that for any row, the sum

$$(abc - cba) + (bca - acb) + (cab - bac) = 0 \quad \text{(B.5)}$$

for all vector terms.

However, bearing in mind that

$$a \wedge b \wedge c = \frac{1}{6} \left[(abc - cba) + (bca - acb) + (cab - bac) \right] \quad \text{(B.6)}$$

or

$$6(a \wedge b \wedge c) = (abc - cba) + (bca - acb) + (cab - bac), \tag{B.7}$$

from Table B.2 we have

$$6(a \wedge b \wedge c) = 6(a_1b_2c_3 - a_1b_3c_2 + a_2b_3c_1 - a_2b_1c_3 + a_3b_1c_2 - a_3b_2c_1)e_{123} \tag{B.8}$$

therefore,

$$a \wedge b \wedge c = (a_1b_2c_3 - a_1b_3c_2 + a_2b_3c_1 - a_2b_1c_3 + a_3b_1c_2 - a_3b_2c_1)e_{123} \tag{B.9}$$

or

$$a \wedge b \wedge c = \begin{vmatrix} a_1 & b_1 & c_1 \\ a_2 & b_2 & c_2 \\ a_3 & b_3 & c_3 \end{vmatrix} e_{123}. \tag{B.10}$$

TABLE B.1   Expansion of the terms in $a \wedge b \wedge c$. For each product, the scalar terms are shown in the left-hand column, with their associated vector in the product column

	abc	cba	bca	acb	cab	bac
	Even	Odd	Even	Odd	Even	Odd
$a_1b_1c_1$	e_1	e_1	e_1	e_1	e_1	e_1
$a_1b_1c_2$	e_2	e_2	$-e_2$	$-e_2$	e_2	e_2
$a_1b_1c_3$	e_3	e_3	$-e_3$	$-e_3$	e_3	e_3
$a_1b_2c_1$	$-e_2$	$-e_2$	e_2	e_2	e_2	e_2
$a_1b_2c_2$	e_1	e_1	e_1	e_1	$-e_1$	$-e_1$
$a_1b_2c_3$	e_{123}	$-e_{123}$	e_{123}	$-e_{123}$	e_{123}	$-e_{123}$
$a_1b_3c_1$	$-e_3$	$-e_3$	e_3	e_3	e_3	e_3
$a_1b_3c_2$	$-e_{123}$	e_{123}	$-e_{123}$	e_{123}	$-e_{123}$	e_{123}
$a_1b_3c_3$	e_1	e_1	e_1	e_1	$-e_1$	$-e_1$
$a_2b_1c_1$	e_2	e_2	e_2	e_2	$-e_2$	$-e_2$
$a_2b_1c_2$	$-e_1$	$-e_1$	e_1	e_1	e_1	e_1
$a_2b_1c_3$	$-e_{123}$	e_{123}	$-e_{123}$	e_{123}	$-e_{123}$	e_{123}
$a_2b_2c_1$	e_1	e_1	$-e_1$	$-e_1$	e_1	e_1
$a_2b_2c_2$	e_2	e_2	e_2	e_2	e_2	e_2
$a_2b_2c_3$	e_3	e_3	$-e_3$	$-e_3$	e_3	e_3
$a_2b_3c_1$	e_{123}	$-e_{123}$	e_{123}	$-e_{123}$	e_{123}	$-e_{123}$
$a_2b_3c_2$	$-e_3$	$-e_3$	e_3	e_3	e_3	e_3
$a_2b_3c_3$	e_2	e_2	e_2	e_2	$-e_2$	$-e_2$
$a_3b_1c_1$	e_3	e_3	e_3	e_3	$-e_3$	$-e_3$
$a_3b_1c_2$	e_{123}	$-e_{123}$	e_{123}	$-e_{123}$	e_{123}	$-e_{123}$
$a_3b_1c_3$	$-e_1$	$-e_1$	e_1	e_1	e_1	e_1
$a_3b_2c_1$	$-e_{123}$	e_{123}	$-e_{123}$	e_{123}	$-e_{123}$	e_{123}
$a_3b_2c_2$	e_3	e_3	e_3	e_3	$-e_3$	$-e_3$
$a_3b_2c_3$	$-e_2$	$-e_2$	e_2	e_2	e_2	e_2
$a_3b_3c_1$	e_1	e_1	$-e_1$	$-e_1$	e_1	e_1
$a_3b_3c_2$	e_2	e_2	$-e_2$	$-e_2$	e_2	e_2
$a_3b_3c_3$	e_3	e_3	e_3	e_3	e_3	e_3

TABLE B.2   Expansion of the terms $a \wedge b \wedge c$ taking into account that odd terms have their signs switched. For each product, the scalar terms are shown in the left-hand column, with their associated vector in the product column. The right-hand column sums each row

	abc	$-cba$	bca	$-acb$	cab	$-bac$	Σ
	Even	Odd	Even	Odd	Even	Odd	
$a_1b_1c_1$	e_1	$-e_1$	e_1	$-e_1$	e_1	$-e_1$	0
$a_1b_1c_2$	e_2	$-e_2$	$-e_2$	e_2	e_2	$-e_2$	0
$a_1b_1c_3$	e_3	$-e_3$	$-e_3$	e_3	e_3	$-e_3$	0
$a_1b_2c_1$	$-e_2$	e_2	e_2	$-e_2$	e_2	$-e_2$	0
$a_1b_2c_2$	e_1	$-e_1$	e_1	$-e_1$	$-e_1$	e_1	0
$a_1b_2c_3$	e_{123}	e_{123}	e_{123}	e_{123}	e_{123}	e_{123}	$6e_{123}$
$a_1b_3c_1$	$-e_3$	e_3	e_3	$-e_3$	e_3	$-e_3$	0
$a_1b_3c_2$	$-e_{123}$	$-e_{123}$	$-e_{123}$	$-e_{123}$	$-e_{123}$	$-e_{123}$	$-6e_{123}$
$a_1b_3c_3$	e_1	$-e_1$	e_1	$-e_1$	$-e_1$	e_1	0
$a_2b_1c_1$	e_2	$-e_2$	e_2	$-e_2$	$-e_2$	e_2	0
$a_2b_1c_2$	$-e_1$	e_1	e_1	$-e_1$	e_1	$-e_1$	0
$a_2b_1c_3$	$-e_{123}$	$-e_{123}$	$-e_{123}$	$-e_{123}$	$-e_{123}$	$-e_{123}$	$-6e_{123}$
$a_2b_2c_1$	e_1	$-e_1$	$-e_1$	e_1	e_1	$-e_1$	0
$a_2b_2c_2$	e_2	$-e_2$	e_2	$-e_2$	e_2	$-e_2$	0
$a_2b_2c_3$	e_3	$-e_3$	$-e_3$	e_3	e_3	$-e_3$	0
$a_2b_3c_1$	e_{123}	e_{123}	e_{123}	e_{123}	e_{123}	e_{123}	$6e_{123}$
$a_2b_3c_2$	$-e_3$	e_3	e_3	$-e_3$	e_3	$-e_3$	0
$a_2b_3c_3$	e_2	$-e_2$	e_2	$-e_2$	$-e_2$	e_2	0
$a_3b_1c_1$	e_3	$-e_3$	e_3	$-e_3$	$-e_3$	e_3	0
$a_3b_1c_2$	e_{123}	e_{123}	e_{123}	e_{123}	e_{123}	e_{123}	$6e_{123}$
$a_3b_1c_3$	$-e_1$	e_1	e_1	$-e_1$	e_1	$-e_1$	0
$a_3b_2c_1$	$-e_{123}$	$-e_{123}$	$-e_{123}$	$-e_{123}$	$-e_{123}$	$-e_{123}$	$-6e_{123}$
$a_3b_2c_2$	e_3	$-e_3$	e_3	$-e_3$	$-e_3$	e_3	0
$a_3b_2c_3$	$-e_2$	e_2	e_2	$-e_2$	e_2	$-e_2$	0
$a_3b_3c_1$	e_1	$-e_1$	$-e_1$	e_1	e_1	$-e_1$	0
$a_3b_3c_2$	e_2	$-e_2$	$-e_2$	e_2	e_2	$-e_2$	0
$a_3b_3c_3$	e_3	$-e_3$	e_3	$-e_3$	e_3	$-e_3$	0

Bibliography

Crowe, M. *A History of Vector Analysis*, Dover Publications, Inc. 1985.
Doran, C. & Lasenby, A. *Geometric Algebra for Physicists*, Cambridge University Press, 2003.
Gullberg, J. *Mathematics from the Birth of Numbers*, W.W. Norton & Co., 1997.
Vince, J. *Geometric Algebra for Computer Graphics*, Springer, 2007.

Index

DATE DUE